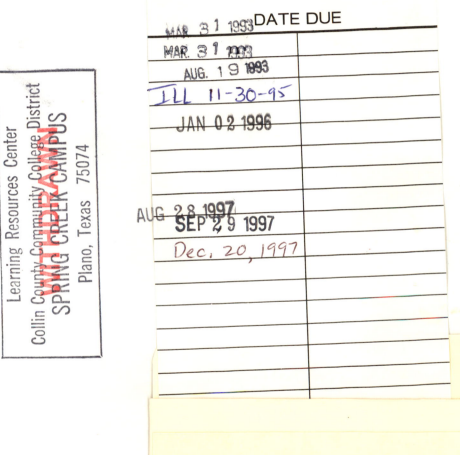

Applications of Linear Algebra

THIRD EDITION

by
Chris Rorres
Howard Anton

DREXEL UNIVERSITY

1979

JOHN WILEY & SONS

New York • Chichester • Brisbane • Toronto • Singapore

ISBN 0-471-86800-0
Printed in the United States of America

10 9 8 7 6 5 4 3 2

To Peter

Preface

This textbook discusses selected applications of linear algebra. The presentation is suitable for students who have completed or are taking concurrently a standard sophomore-level course in linear algebra.

Topics are drawn from a wide variety of fields including business, economics, engineering, physics, computer science, geometry, approximation theory, ecology, sociology, demography, and genetics. Also included is a brief introduction to game theory, Markov chains, and graph theory. At the end of the text there is a three chapter minicourse in linear programming which can be covered in about six lectures.

With a few clearly-marked exceptions, each application is in its own *independent* chapter, so that chapters can be deleted or permuted freely to fit individual needs and interests. Each topic begins with a list of linear algebra prerequisites in order that a reader can tell in advance if he or she has sufficient background to read the chapter.

Since the topics vary considerably in difficulty, we have included a *subjective* rating of each topic — easy, moderate, more difficult. (See the *Guide for the Instructor* following this preface.) Our evaluation is based more on the intrinsic difficulty of the material rather than the number of prerequisites; thus a topic requiring fewer mathematical prerequisites may be rated harder than one requiring more prerequisites.

Since our primary objective is to present applications of linear algebra, proofs are often omitted. We assume the reader has met the linear algebra prerequisites and whenever results from other fields are needed, they are stated precisely (with motivation where possible), but usually without proof.

Although this text was written to be used with Howard Anton's *Elementary Linear Algebra*, John Wiley and Sons, Inc., we have avoided specialized notation or terminology so that this book can be utilized in conjunction with any standard undergraduate text in linear algebra. gebra.

There are several possible ways to use this book:

(a) as a supplement to a standard linear algebra text,
(b) as a textbook for a follow-up course to linear algebra,
(c) as part of a self-study enrichment program or an introduction to mathematical research.

In addition, this text may serve as a source of topics for a mathematical modeling or computer programming course.

Three new chapters are included in this third edition:

- Cubic Spline Interpolation
- Cryptography
- Computed Tomography

A solutions manual is available for the third edition. It contains complete solutions to all of the exercises in the text.

We would like to express our appreciation to Kathleen R. McCabe of Techni-Type who typed the entire manuscript. Her patience and skill contributed greatly to the appearance of this text. Our thanks are also due to Dennis DeTurck, Charles Shuman, and Miriam Siegler, who assisted with the solutions manual, exercises, and examples. We also express our appreciation for the guidance provided by the reviewers: Keith J. Craswell of Western Washington University, Harry W. McLaughlin of Rensselaer Polytechnic Institute, Roberto Mena of the University of Wyoming, William F. Trench of Drexel University, and Oleh J. Tretiak of Drexel University. Our thanks also to the Medical Systems Division of General Electric Company for the photographs and diagrams they provided for the Computed Tomography chapter. Finally, we thank our editor, Robert Pirtle, our production supervisor, Jennifer Salerno, and the entire Wiley staff for their encouragement and guidance.

Chris Rorres

Howard Anton

Contents

Guide for the Instructor

The table below classifies the first sixteen chapters according to difficulty as follows:

Easy —— *the average student who has met the stated prerequisites should be able to read the material with no help from the instructor.*

Moderate —— *the average student who has met the stated prerequisites may require a little help from the instructor.*

More Difficult —— *the average student who has met the stated prerequisites will probably need help from the instructor.*

	Chapter																		
	1	2	3	4	5	6	7	8	9	10	11	12	13	14	15	16	17	18	19
Easy	●	●													●				
Moderate				●	●	●	●	●	●	●	●					●		●	
More Difficult			●									●	●	●			●		●

With the exception of Chapter 14, which depends on Chapter 13, the first sixteen chapters are independent and can be presented in any order. Chapters 20, 21, and 22 provide an introduction to linear programming suitable for students at the sophomore level. This material does not depend on the rest of the text and can be presented as soon as the students have met the linear algebra prerequisites.

Constructing Curves and Surfaces through Specified Points

A technique for using determinants to construct lines, circles, and general conic sections through specified points in the plane is described. The procedure is also used to pass planes and spheres in three-dimensional space through fixed points.

PREREQUISITES: Linear systems
                        Determinants
                        Analytic geometry

INTRODUCTION

One of the fundamental results in the theory of Linear Algebra is the following:

> *A homogeneous linear system with as many equations as unknowns has a nontrivial solution if and only if the determinant of the system is zero.*

In this chapter, we show how this result may be used to determine the equations of various curves and surfaces through specified points. We proceed immediately to some specific examples.

A Line Through Two Points

Suppose we are given two distinct points in the plane, (x_1, y_1) and (x_2, y_2). There is a unique line,

$$c_1 x + c_2 y + c_3 = 0, \qquad (1.1)$$

which passes through these two points. Notice that c_1, c_2, and c_3 are not all zero, and that these coefficients are unique only up to a multiplicative constant.

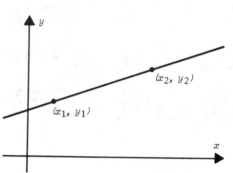

Since (x_1, y_1) and (x_2, y_2) lie on the line, substituting them in (1.1) gives the two equations

$$c_1 x_1 + c_2 y_1 + c_3 = 0 \qquad (1.2)$$

$$c_1 x_2 + c_2 y_2 + c_3 = 0. \qquad (1.3)$$

The three equations, (1.1), (1.2), and (1.3), may be grouped together in system form as

$$x\, c_1 + y\, c_2 + c_3 = 0$$
$$x_1 c_1 + y_1 c_2 + c_3 = 0$$
$$x_2 c_1 + y_2 c_2 + c_3 = 0.$$

In this form, we have a homogeneous system of three equations for c_1, c_2, and c_3. Since c_1, c_2, and c_3 are not all zero, this system has a nontrivial solution, and so the determinant of the system must be zero. That is,

$$\begin{vmatrix} x & y & 1 \\ x_1 & y_1 & 1 \\ x_2 & y_2 & 1 \end{vmatrix} = 0. \qquad (1.4)$$

Consequently, every point (x, y) on the line satisfies (1.4), and conversely, every point (x, y) which satisfies (1.4) lies on the line.

EXAMPLE 1.1 Find the equation of the line which passes through the two points $(2, 1)$ and $(3, 7)$.

SOLUTION Substitution of the coordinates of the two points into Eq. (1.4) gives

$$\begin{vmatrix} x & y & 1 \\ 2 & 1 & 1 \\ 3 & 7 & 1 \end{vmatrix} = 0.$$

The cofactor expansion of this determinant along the first row then gives:

$$-6x + y + 11 = 0.$$

A CIRCLE THROUGH THREE POINTS

Let us be given three distinct points in the plane, (x_1, y_1), (x_2, y_2), and (x_3, y_3), not all lying on a straight line. From analytic geometry, we know that there is a unique circle, say

$$c_1(x^2 + y^2) + c_2 x + c_3 y + c_4 = 0, \quad (1.5)$$

which passes through them (Fig. 1.2). Substituting the coordinates of the three points into this equation gives

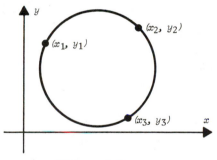

Figure 1.2

$$c_1(x_1^2 + y_1^2) + c_2 x_1 + c_3 y_1 + c_4 = 0 \qquad (1.6)$$

$$c_1(x_2^2 + y_2^2) + c_2 x_2 + c_3 y_2 + c_4 = 0 \qquad (1.7)$$

$$c_1(x_3^2 + y_3^2) + c_2 x_3 + c_3 y_3 + c_4 = 0. \qquad (1.8)$$

As before, Eqs. (1.5) - (1.8) form a homogeneous linear system with a nontrivial solution for c_1, c_2, c_3, and c_4. Thus the determinant of this linear system is zero:

$$\begin{vmatrix} x^2 + y^2 & x & y & 1 \\ x_1^2 + y_1^2 & x_1 & y_1 & 1 \\ x_2^2 + y_2^2 & x_2 & y_2 & 1 \\ x_3^2 + y_3^2 & x_3 & y_3 & 1 \end{vmatrix} = 0. \qquad (1.9)$$

This is a determinant form for the equation of the circle.

EXAMPLE 1.2 Find the equation of the circle which passes through the three points $(1, 7)$, $(6, 2)$, and $(4, 6)$.

SOLUTION Substitution of the coordinates of the three points into Eq. (1.9) gives

$$\begin{vmatrix} x^2 + y^2 & x & y & 1 \\ 50 & 1 & 7 & 1 \\ 40 & 6 & 2 & 1 \\ 52 & 4 & 6 & 1 \end{vmatrix} = 0,$$

which reduces to

$$10(x^2 + y^2) - 20x - 40y - 200 = 0.$$

In standard form this is

$$(x - 1)^2 + (y - 2)^2 = 5^2.$$

Thus, the circle has center $(1, 2)$ and radius 5.

A GENERAL CONIC SECTION THROUGH FIVE POINTS

The general equation of a conic section in the plane (a parabola, hyperbola, or ellipse, and degenerate forms of these three curves) is given by

$$c_1 x^2 + c_2 xy + c_3 y^2 + c_4 x + c_5 y + c_6 = 0.$$

Figure 1.3

This form contains six coefficients, although only five are needed since we may divide through by any one of them which is not zero. Thus, only five coefficients must be determined, so that five distinct points in the plane are sufficient to determine the equation of the conic section (Fig. 1.3). As before, the equation may be put in determinant form (see Exercise 1.6):

$$
\begin{vmatrix}
x^2 & xy & y^2 & x & y & 1 \\
x_1^2 & x_1 y_1 & y_1^2 & x_1 & y_1 & 1 \\
x_2^2 & x_2 y_2 & y_2^2 & x_2 & y_2 & 1 \\
x_3^2 & x_3 y_3 & y_3^2 & x_3 & y_3 & 1 \\
x_4^2 & x_4 y_4 & y_4^2 & x_4 & y_4 & 1 \\
x_5^2 & x_5 y_5 & y_5^2 & x_5 & y_5 & 1
\end{vmatrix} = 0.
\tag{1.10}
$$

EXAMPLE 1.3   An astronomer wants to determine the orbit of an asteroid about the sun. He sets up a Cartesian coordinate system in the plane of the orbit with the sun at the origin. Astronomical units of measurement are used along the axes. (1 astronomical unit = mean distance of earth to sun = 93 million miles.) By Kepler's first law, he knows that the orbit must be an ellipse. Consequently, he makes five observations of the asteroid at five different times and finds five points along the orbit to be

$(5.764, 0.648)$, $(6.286, 1.202)$, $(6.759, 1.823)$,

$(7.168, 2.526)$, $(7.480, 3.360)$.

Find the equation of the orbit.

SOLUTION   Substitution of the coordinates of the five given points into Eq. (1.10) gives

$$
\begin{vmatrix}
x^2 & xy & y^2 & x & y & 1 \\
33.224 & 3.735 & 0.420 & 5.764 & 0.648 & 1 \\
39.514 & 7.556 & 1.445 & 6.286 & 1.202 & 1 \\
45.684 & 12.322 & 3.323 & 6.759 & 1.823 & 1 \\
51.380 & 18.106 & 6.381 & 7.168 & 2.526 & 1 \\
55.950 & 25.133 & 11.290 & 7.480 & 3.360 & 1
\end{vmatrix} = 0.
$$

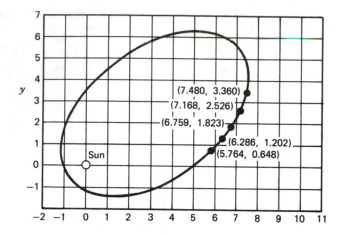

Figure 1.4

The cofactor expansion of this determinant along the first row reduces to

$$x^2 - 1.04xy + 1.30y^2 - 3.90x - 2.93y - 5.49 = 0.$$

Figure 1.4 is a diagram of the orbit, together with the five given points.

A PLANE THROUGH THREE POINTS

In Exercise 1.7 we ask the reader to show the following: The plane in 3-space with equation

$$c_1 x + c_2 y + c_3 z + c_4 = 0$$

which passes through three noncollinear points (x_1, y_1, z_1), (x_2, y_2, z_2), and (x_3, y_3, z_3) is given by the determinant equation

$$\begin{vmatrix} x & y & z & 1 \\ x_1 & y_1 & z_1 & 1 \\ x_2 & y_2 & z_2 & 1 \\ x_3 & y_3 & z_3 & 1 \end{vmatrix} = 0. \tag{1.11}$$

EXAMPLE 1.4   The equation of the plane which passes through the three noncollinear points $(1, 1, 0)$, $(2, 0, -1)$, and $(2, 9, 2)$ is

$$\begin{vmatrix} x & y & z & 1 \\ 1 & 1 & 0 & 1 \\ 2 & 0 & -1 & 1 \\ 2 & 9 & 2 & 1 \end{vmatrix} = 0,$$

which reduces to

$$2x - y + 3z - 1 = 0.$$

A Sphere Through Four Points

In Exercise 1.8 we ask the reader to show the following: The sphere in 3-space with equation

$$c_1(x^2 + y^2 + z^2) + c_2 x + c_3 y + c_4 z + c_5 = 0$$

which passes through four noncoplanar points (x_1, y_1, z_1), (x_2, y_2, z_2), (x_3, y_3, z_3), and (x_4, y_4, z_4) is given by the following determinant equation:

$$\begin{vmatrix} x^2 + y^2 + z^2 & x & y & z & 1 \\ x_1^2 + y_1^2 + z_1^2 & x_1 & y_1 & z_1 & 1 \\ x_2^2 + y_2^2 + z_2^2 & x_2 & y_2 & z_2 & 1 \\ x_3^2 + y_3^2 + z_3^2 & x_3 & y_3 & z_3 & 1 \\ x_4^2 + y_4^2 + z_4^2 & x_4 & y_4 & z_4 & 1 \end{vmatrix} = 0. \qquad (1.12)$$

EXAMPLE 1.5 The equation of the sphere which passes through the four points $(0, 3, 2)$, $(1, -1, 1)$, $(2, 1, 0)$, and $(5, 1, 3)$ is

$$\begin{vmatrix} x^2 + y^2 + z^2 & x & y & z & 1 \\ 13 & 0 & 3 & 2 & 1 \\ 3 & 1 & -1 & 1 & 1 \\ 5 & 2 & 1 & 0 & 1 \\ 35 & 5 & 1 & 3 & 1 \end{vmatrix} = 0.$$

This reduces to

$$x^2 + y^2 + z^2 - 4x - 2y - 6z + 5 = 0,$$

which in standard form is

$$(x - 2)^2 + (y - 1)^2 + (z - 3)^2 = 9.$$

EXERCISES

1.1 Find the equations of the lines which pass through the following points:

(a) $(1,-1)$, $(2,2)$
(b) $(0,1)$, $(1,-1)$.

1.2 Find the equations of the circles which pass through the following points:

(a) $(2, 6)$, $(2,0)$, $(5,3)$
(b) $(2,-2)$, $(3,5)$, $(-4,6)$.

1.3 Find the equation of the conic section which passes through the points $(0,0)$, $(0,-1)$, $(2,0)$, $(2,-5)$, and $(4,-1)$.

1.4 Find the equations of the planes in 3-space which pass through the following points:

(a) $(1,1,-3)$, $(1,-1,1)$, $(0,-1,2)$
(b) $(2,3,1)$, $(2,-1,-1)$, $(1,2,1)$.

1.5 Find the equations of the spheres in 3-space which pass through the following points:

(a) $(1,2,3)$, $(-1,2,1)$, $(1,0,1)$, $(1,2,-1)$
(b) $(0,1,-2)$ $(1,3,1)$ $(2,-1,0)$ $(3,1,-1)$.

1.6 Show that Eq. (1.10) is the equation of the conic section which passes through five given distinct points.

1.7 Show that Eq. (1.11) is the equation of the plane in 3-space which passes through three given noncollinear points.

1.8 Show that Eq. (1.12) is the equation of the sphere in 3-space which passes through four given noncoplanar points.

1.9 Find a determinant equation for the parabola of the form

$$c_1 y + c_2 x^2 + c_3 x + c_4 = 0$$

which passes through three given noncollinear points in the plane.

Plane Geometry

2

Vector algebra is used to prove theorems in plane geometry.

PREREQUISITES: Plane geometry
                  Vector algebra
                  Dot product

INTRODUCTION

The basic properties of Euclidean space were first systematically studied 23 centuries ago by Euclid, who began with accepted axioms and deduced theorems in plane and solid geometry from them. In this chapter, we prove some elementary theorems in plane geometry using vector algebra. This method usually leads to simpler and more elegant proofs than the methods studied in a first course in geometry — the same methods that Euclid himself employed. This is because the basic equations of vector algebra, such as

$$a + b = b + a$$
$$a + (b + c) = (a + b) + c$$
$$a + (-a) = 0$$
$$\vdots$$

9

are vector space axioms and therefore apply to Euclidean space. Thus, any new equation we can generate from these basic equations constitutes a theorem. Since it is easier to manipulate equations than to write out in words how one is applying axioms, a certain economy of thought is attained and a cleaner presentation of a proof results.

Before presenting some examples of such vector techniques applied to plane geometry, we introduce some notation and review some basic definitions.

NOTATION AND DEFINITIONS

For our applications to plane geometry, a vector may be viewed as a directed line segment. We shall use the following notation (Fig. 2.1): If A and B are two points, then AB denotes the (undirected) line segment connecting A and B and $\overrightarrow{AB}$ denotes the directed line segment or vector from point A to point B. We shall use **a** and **b** to denote vectors from an arbitrary fixed reference point 0 to the points A and B, respectively. Notice that we may write

Figure 2.1

$$\overrightarrow{AB} = \mathbf{b} - \mathbf{a}.$$

We assume that the reader is familiar with the usual rules of vector algebra. However, it will be convenient to state the following criterion for determining when three points are collinear:

Let A, B, and C be three points and let **a**, **b**, *and* **c** *be the vectors from a fixed point 0 to A, B, and C, respectively (Fig. 2.2). Then B lies on the line segment AC if and only if*

$$\mathbf{b} = \lambda \mathbf{a} + (1 - \lambda)\mathbf{c}$$

for some number λ satisfying

$$0 \leq \lambda \leq 1.$$

Figure 2.2

When the above criterion is satisfied, the point B cuts the line segment AC in the ratio $\lambda:(1-\lambda)$. For example, if $\lambda = \frac{1}{2}$, i.e. if $b = \frac{1}{2}(a+c)$, then B is the midpoint of AC; while if $\lambda = \frac{2}{3}$, i.e. if $b = \frac{2}{3}a + \frac{1}{3}c$, then B cuts AC in the ratio 2:1.

We also recall the definition of the dot (or scalar or inner) product $a \cdot b$ of two nonzero vectors a and b:

$$a \cdot b = \|a\| \|b\| \cos \theta$$
$$= ab \cos \theta$$

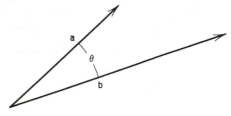

Here, $\|a\|$ or a is the length of vector a, $\|b\|$ or b is the length of vector b, and θ is the angle between a and b (Fig. 2.3). A useful property of the dot product is the fact that two nonzero vectors are perpendicular to each other if and only if their dot product is zero.

Figure 2.3

In Fig. 2.4 on the next page, seven definitions from plane geometry are listed for reference. These are followed by six examples in which six basic theorems are proved using vector algebra.

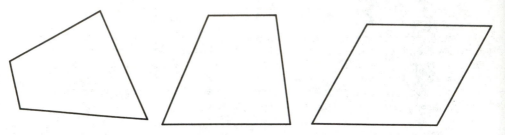

A **quadrilateral** is a four-sided polygon.

A **trapezoid** is a quadrilateral with one pair of parallel sides.

A **parallelogram** is a quadrilateral with two pairs of parallel sides.

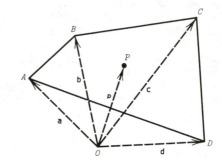

The **centroid or mean center of a triangle** ABC is the point P determined by

$$\mathbf{p} = \frac{1}{3}(\mathbf{a} + \mathbf{b} + \mathbf{c}).$$

The **centroid or mean center of a quadrilateral** $ABCD$ is the point P determined by

$$\mathbf{p} = \frac{1}{4}(\mathbf{a} + \mathbf{b} + \mathbf{c} + \mathbf{d}).$$

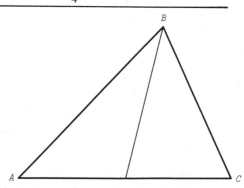

An **altitude** of a triangle ABC is a line segment from one of the vertices perpendicular to the opposite side.

A **median** of a triangle ABC is a line segment from one of the vertices to the midpoint of the opposite side.

Figure 2.4

EXAMPLE 2.1

Prove that the three medians of a triangle intersect at the centroid of the triangle.

Figure 2.5

PROOF. If D, E, and F are the midpoints of the three sides of the triangle ABC, we have

$$d = \tfrac{1}{2}(a + c)$$
$$e = \tfrac{1}{2}(a + b)$$
$$f = \tfrac{1}{2}(b + c).$$

The centroid P of the triangle is determined by

$$p = \frac{1}{3}(a + b + c).$$

Substituting $a + c = 2d$ into this last equation gives

$$p = \frac{1}{3}b + \frac{2}{3}d,$$

from which it follows that P lies on the median BD. Similarly, the equations

$$p = \frac{1}{3}c + \frac{2}{3}e$$
$$p = \frac{1}{3}a + \frac{2}{3}f$$

show that P also lies on the medians CE and AF.

Remark: Notice that from these last three equations for p it follows that the centroid divides each median in the ratio 2:1.

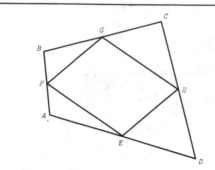

Figure 2.6

EXAMPLE 2.2

Prove that the quadrilateral *EFGH* whose vertices are the midpoints of the sides of an arbitrary quadrilateral *ABCD* is a parallelogram. In addition, prove that the two quadrilaterals have the same centroid.

PROOF. We have that

$$e = \tfrac{1}{2}(a + d)$$
$$f = \tfrac{1}{2}(a + b)$$
$$g = \tfrac{1}{2}(b + c)$$
$$h = \tfrac{1}{2}(c + d).$$

Thus $g - f = \tfrac{1}{2}(c - a) = h - e$, and so *FG* is parallel to *EH*. Similarly, $h - g = \tfrac{1}{2}(d - b) = e - f$, and so *GH* is parallel to *FE*. Thus quadrilateral *EFGH* is a parallelogram.

Adding the above four equations yields

$$e + f + g + h = a + b + c + d$$

or

$$\frac{1}{4}(e + f + g + h) = \frac{1}{4}(a + b + c + d),$$

and so both quadrilaterals have the same centroid.

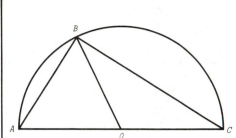

EXAMPLE 2.3

Prove that an angle inscribed in
a semicircle is a right angle.

Figure 2.7

PROOF, We want to show that the vector $\vec{AB}$ is perpendicular to
the vector $\vec{BC}$. Letting O be the center of the semicircle, we
have

$$\vec{AB} \cdot \vec{BC} = (\vec{AO} + \vec{OB}) \cdot (\vec{BO} + \vec{OC})$$

$$= (\vec{OC} + \vec{OB}) \cdot (-\vec{OB} + \vec{OC})$$

$$= \vec{OC} \cdot \vec{OC} - \vec{OB} \cdot \vec{OB}$$

$$= \|\vec{OC}\|^2 - \|\vec{OB}\|^2$$

$$= 0$$

where the last equality follows since $\vec{OC}$ and $\vec{OB}$ have the same
length.

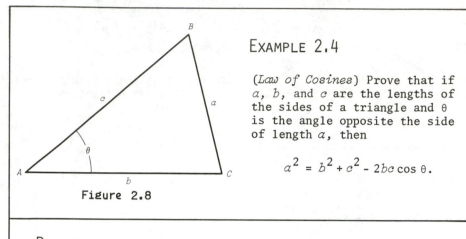

Figure 2.8

EXAMPLE 2.4

(*Law of Cosines*) Prove that if a, b, and c are the lengths of the sides of a triangle and θ is the angle opposite the side of length a, then

$$a^2 = b^2 + c^2 - 2bc \cos \theta.$$

PROOF. Letting A, B, and C be the vertices opposite the sides of lengths a, b, and c, respectively, we have

$$
\begin{aligned}
a^2 &= \|\overrightarrow{BC}\|^2 \\
&= \|\overrightarrow{BA} + \overrightarrow{AC}\|^2 \\
&= (\overrightarrow{BA} + \overrightarrow{AC}) \cdot (\overrightarrow{BA} + \overrightarrow{AC}) \\
&= \overrightarrow{AC} \cdot \overrightarrow{AC} + \overrightarrow{BA} \cdot \overrightarrow{BA} + 2(\overrightarrow{AC} \cdot \overrightarrow{BA}) \\
&= \overrightarrow{AC} \cdot \overrightarrow{AC} + \overrightarrow{BA} \cdot \overrightarrow{BA} - 2(\overrightarrow{AC} \cdot \overrightarrow{AB}) \\
&= \|\overrightarrow{AC}\|^2 + \|\overrightarrow{BA}\|^2 - 2\|\overrightarrow{AC}\| \|\overrightarrow{AB}\| \cos \theta \\
&= b^2 + c^2 - 2bc \cos \theta.
\end{aligned}
$$

Figure 2.9

EXAMPLE 2.5

Prove that the three altitudes of a triangle intersect at a common point (called the *orthocenter* of the triangle).

PROOF: Let R be the point of intersection of the altitudes from vertices A and B, and let CE be the line from vertex C passing through R and meeting side AB at point E. We want to show that CE is perpendicular to AB.

We have that

$$\vec{CA} \cdot \vec{BR} = (a - c) \cdot (r - b) = 0$$

and

$$\vec{CB} \cdot \vec{AR} = (b - c) \cdot (r - a) = 0.$$

Thus

$$
\begin{aligned}
\vec{AB} \cdot \vec{CR} &= (b - a) \cdot (r - c) \\
&= [(b - c) - (a - c)] \cdot (r - c) \\
&= (b - c) \cdot (r - c) - (a - c) \cdot (r - c) \\
&= (b - c) \cdot (r - a + a - c) - (a - c) \cdot (r - b + b - c) \\
&= (b - c) \cdot (r - a) + (b - c) \cdot (a - c) \\
&\quad - (a - c) \cdot (r - b) - (a - c) \cdot (b - c) \\
&= (b - c) \cdot (a - c) - (a - c) \cdot (b - c) \\
&= 0,
\end{aligned}
$$

which proves that CR, and hence CE, is perpendicular to AB.

EXAMPLE 2.6

Prove that the three perpendicular bisectors of the sides of a
triangle intersect at a common
point (called the *circumcenter* of
the triangle).

Figure 2.10

PROOF. Let Q be the point of intersection of the perpendicular
bisectors of sides AB and AC, and let F be the midpoint of side
BC. We want to show that FQ is perpendicular to CB.

We have that

$$\overrightarrow{AB} \cdot \overrightarrow{EQ} = (b - a) \cdot (q - e) = 0$$

and

$$\overrightarrow{CA} \cdot \overrightarrow{DQ} = (a - c) \cdot (q - d) = 0.$$

Thus

$$
\begin{aligned}
\overrightarrow{CB} \cdot \overrightarrow{FQ} &= (b - c) \cdot (q - f) \\
&= [(b - a) + (a - c)] \cdot (q - f) \\
&= (b - a) \cdot (q - f) + (a - c) \cdot (q - f) \\
&= (b - a) \cdot [q - \tfrac{1}{2}(b + c)] + (a - c) \cdot [q - \tfrac{1}{2}(b + c)] \\
&= (b - a) \cdot [q - \tfrac{1}{2}(a + b) + \tfrac{1}{2}(a - c)] \\
&\quad + (a - c) \cdot [q - \tfrac{1}{2}(a + c) + \tfrac{1}{2}(a - b)] \\
&= (b - a) \cdot [q - e + \tfrac{1}{2}(a - c)] + (a - c) \cdot [q - d + \tfrac{1}{2}(a - b)] \\
&= \tfrac{1}{2}(b - a) \cdot (a - c) + \tfrac{1}{2}(a - c) \cdot (a - b) \\
&= 0,
\end{aligned}
$$

which proves that FQ is perpendicular to CB.

EXERCISES

2.1 Let D and E be the midpoints of the sides AB and BC of a triangle ABC. Prove that DE is parallel to side AC and that its length is one-half that of AC.

Figure 2.11

2.2 Prove that the line segments FH and EG connecting opposite midpoints of the sides of a quadrilateral $ABCD$ intersect at the centroid P of the quadrilateral.

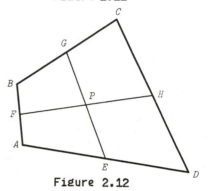

Figure 2.12

2.3 Prove that the diagonals AC and BD of a parallelogram $ABCD$ bisect each other and that the point of bisection P is the centroid of the parallelogram.

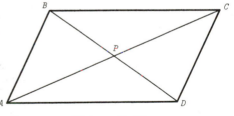

Figure 2.13

2.4 Prove that the circumcenter Q of a triangle ABC is equidistant from its three vertices, and hence is the center of the circle which circumscribes the triangle. (See Example 2.6.)

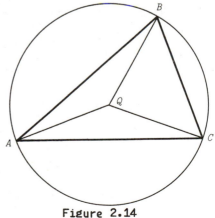

Figure 2.14

2.5 Prove that the centroid P of a triangle ABC lies on the line segment QR connecting the circumcenter Q of the triangle to its orthocenter R. In addition, prove that P divides this line (called the *Euler Line*) in the ratio 2:1. (See Examples 2.5 and 2.6.)

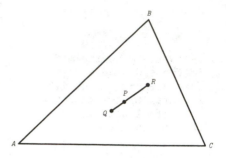

Figure 2.15

2.6 Let E and F be points on the sides AB and BC of a triangle ABC such that EF is parallel to side AC. Prove that if D is the point on side AC such that BD passes through the intersection G of EC and AF, then D bisects AC.

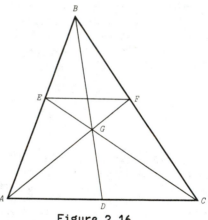

Figure 2.16

2.7 Let E be the midpoint of the side CD of a parallelogram $ABCD$. Prove that the point of intersection F of BE and diagonal AC divides AC in the ratio 2:1.

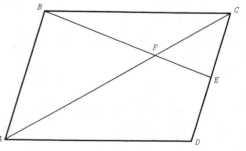

Figure 2.17

2.8 Prove that the diagonals *AC* and *BD* of a trapezoid *ABCD* cut each other in the same ratio.

Figure 2.18

3

Equilibrium of Rigid Bodies

Vector algebra is applied to the analysis of the equilibrium of two-dimensional rigid bodies.

PREREQUISITES: Vector Algebra

Introduction

The field of mechanics greatly stimulated the mathematical study of vectors and vector spaces. In this chapter, we examine the use of vectors in that branch of static mechanics concerned with the equilibrium of rigid bodies. To simplify the presentation, we shall assume that the forces acting on the bodies all lie in a common plane, though the bodies themselves may be three dimensional.

Forces and Moments

We will be concerned with the forces acting on rigid bodies. These forces may arise from a variety of sources, such as applied loads, gravity, and reactions of walls, floors, supports, cables, pins, or other bodies in contact with the given body. A force, whatever its nature, is specified by its magnitude, direction, and point of application. Mathematically, a force is an example of a

vector. Consequently, a force
can be represented graphically
by an arrow or directed line
segment (Fig. 3.1) whose length
is proportional to the magnitude
of the force, whose tail is at
the point of application of the
force, and whose direction is
the same as that of the force.
The line passing through the
point of application of a force
and lying along its direction is
called the *line of action* of the
force. For a rigid body, the
following principle is of basic
importance:

Figure 3.1

Principle of Transmissibility

*The effect of a force on a rigid body is not changed
if the force vector is moved along its line of action.*

(a)

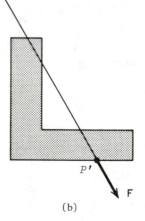

(b)

Figure 3.2

Thus the force F applied at point P of the L-shaped rigid body in Fig. 3.2(a) has the same effect as the same force F applied at point P' in Fig. 3.2(b) since both forces are equal as vectors and have the same lines of action. It is this principle of transmissibility which distinguishes a rigid body from a deformable body. Forces which obey this principle are examples of *line* or *sliding vectors*.

Another basic principle of mechanics we shall need is the following:

Resultant of Forces

The effect of any number of forces applied at a single point of a rigid body is not changed if they are replaced by their vector sum applied at that same point. This vector sum is called the **resultant** *of the forces.*

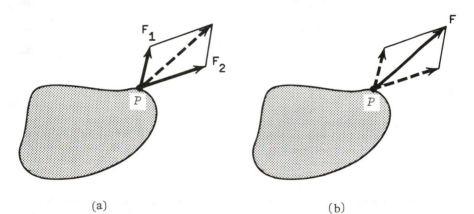

(a)   (b)

Figure 3.3

For example, the two forces F_1 and F_2 applied at point P of the rigid body in Fig. 3.3(a) can be replaced by the single force F in Fig. 3.3(b) obtained by the usual parallelogram law of vector addition. In the same way, a single force may be decomposed vectorially into several components at its point of application without altering its effect on the body.

From Newton's first law of motion, a necessary condition that a body be in equilibrium is that the vector sum of all forces acting on it be zero. If the forces are concurrent (i.e. if their lines of action all pass through a common point) then this condition is also sufficient to guarantee equilibrium since then all the forces can be replaced by their zero resultant at the concurrent point. However, if the forces are not concurrent, then they may impart a rotational motion to the body even though their vector sum is zero. To determine an additional equilibrium condition to prevent rotational motion we must examine the tendency of a force to rotate a body about a point. This leads to the concept of the *moment of a force*.

Moment of a Force

In the plane, the moment M_0 of a force F about a point 0 is the product of the magnitude F of the force and the distance d from the point 0 to the line of action of the force:

$$M_0 = Fd.$$

By convention, the algebraic sign of M_0 is positive if the force F tends to rotate a body counterclockwise about 0, and negative if it tends to rotate a body clockwise about 0.

Figure 3.4

If there are several forces acting on a body, then the individual moments about a fixed point are added (with proper attention to their algebraic signs) to produce the total moment of the forces on the body about the point. The condition of equilibrium needed to prevent rotational motion is that the total moment about some fixed point is zero. The point about which the moments are computed is not important since it can be shown that if the vector sum of the forces is zero and if the sum of their moments about any one point is zero, then the sum of their moments is zero about all points.

Before discussing these conditions of equilibrium in detail, we mention the following theorem of mechanics which simplifies the calculation of moments:

Varignon's Theorem

Let **F** *be the resultant of the two forces* **F₁** *and* **F₂**. *Then the moment of* **F** *about any point 0 is equal to the sum of the moments of* **F₁** *and* **F₂** *about that same point.*

For example, referring to Fig. 3.5, Varignon's Theorem states that

$$Fd = F_1 d_1 + F_2 d_2.$$

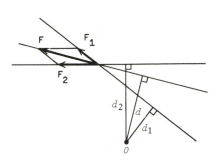

Figure 3.5

EQUILIBRIUM OF RIGID BODIES

From the discussion above, the following two conditions assure that a rigid body in the plane is in equilibrium (more precisely, these conditions serve as the definition of equilibrium):

Equilibrium of Rigid Bodies in the Plane

A rigid body acted upon by coplanar forces is in equilibrium if

1. *the vector sum of the forces is zero, and*

2. *the sum of the moments of the forces about any one point in the plane is zero.*

In condition 2, any point in the plane can be used; however, a judicious choice will considerably reduce the calculations of unknown forces in most problems.

Before presenting some examples, we introduce the idea of a *free-body diagram*. This is a diagram of an isolated rigid body that graphically contains all external forces acting on it. For example, consider the 10-pound slender rod in Fig. 3.6(a) which is

pinned to a wall at one end and
whose other end leans against a
smooth, frictionless wall. Figure
3.6(b) is a free-body diagram of
this rod. The weight of the body
is indicated by a 10-pound force
acting downward through its center
of gravity. For a uniform rod,
this is at its geometric center.
The reaction of the smooth wall is
a force N of unknown magnitude
whose direction is perpendicular
to the wall. The reaction of the
pin is a force F of unknown magnitude and direction.

Another example is the 5-pound
rod in Fig. 3.6(c) which is held
vertically against a rough floor
by a cable. In the free-body diagram of this rod (Fig. 3.6(d)),
the weight is the 5-pound force
acting downward through its geometric center. The cable can
exert a force only along its
length, and so is replaced by a
force T which has the same direction as the cable. The rough
floor can exert both a normal
force and a frictional force parallel to its surface; consequently,
the reaction of the floor is a
force R of unknown magnitude and
direction.

The general technique of
applying the conditions of equilibrium to determine unknown forces
is illustrated in the following
four examples. In these examples,
the horizontal scalar component of
a force will be considered positive if that component is directed
to the right, and negative if it
is directed to the left. Similarly,
vertical scalar components will be
considered positive if directed
upward and negative if directed
downward.

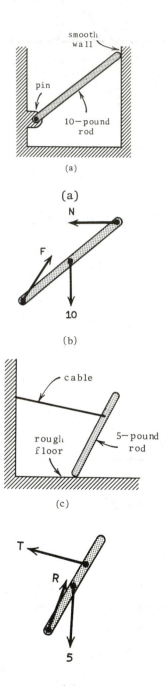

(a)

(a)

(b)

(c)

(d)

Figure 3.6

Figure 3.7

EXAMPLE 3.1 A 1000-pound piano is being raised by a cable. A man guides the piano with a second cable tied to the first. Find the tension in the cable the man pulls on when the piano is in equilibrium as in Fig. 3.7.

Figure 3.8

SOLUTION. We construct a free-body diagram of the common point of the two cables as in Fig. 3.8. Adding the scalar components of the forces in the horizontal direction and setting this sum equal to zero gives the equation

$$-F \sin 5° + T \cos 50° = 0.$$

Similarly, equating the sum of the vertical scalar components equal to zero gives

$$F \cos 5° - 1000 - T \sin 50° = 0.$$

Eliminating F between these two equations and solving for T then yields

$$T = \frac{1000}{\cot 5° \cos 50° - \sin 50°} = 152 \text{ pounds.}$$

Figure 3.9

EXAMPLE 3.2 A 10-foot ladder weighing 30 pounds leans against a smooth vertical wall and makes an angle of 65° with a rough floor. What is the magnitude of the normal force of the wall against the ladder?

Figure 3.10

SOLUTION: In the free-body diagram of the ladder (Fig. 3.10), the weight (30 pounds) acts vertically downward through the geometric center of the ladder, the reaction N of the smooth wall acts horizontally (perpendicular to the wall), and the force F at the foot of the ladder O acts in some unspecified direction. Taking moments about O and equating their sum to zero gives

$$-30(5 \cos 65°) + N(10 \sin 65°) = 0$$

or

$$N = 15 \cot 65° = 7.0 \text{ pounds.}$$

Remark: Notice that by taking moments about O, the unknown force F does not appear in the moment equation. Once N has been found this way, the force F can be found using the equilibrium condition that the sum of all forces is zero.

EXAMPLE 3.3 An 8-ounce spool of
thread is to be lifted over a
1/2-inch step by pulling horizontally on the thread. What force
is necessary to just start it over
the step?

Figure 3.11

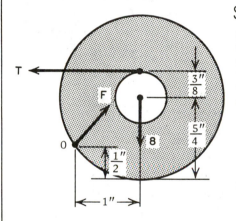

SOLUTION. In the free-body diagram of the spool (Fig. 3.12), the
only reactive force on the spool
is the force F at the edge O of
the step. Taking moments about
this point gives the equilibrium
equation

$$T\left(\frac{3}{8} + \frac{5}{4} - \frac{1}{2}\right) - 8(1) = 0,$$

from which it follows that

$$T = 7\frac{1}{9} \text{ ounces.}$$

Figure 3.12

Remark: Notice that by taking moments about the point of
application of the unknown force F, that unknown
force does not appear in the moment equation.
Once the tension force T is computed, F can be
found by the equilibrium condition that the sum
of all forces is zero.

Figure 3.13

EXAMPLE 3.4 A 200-pound sign is
supported on a pole by two pins
and a cable. What is the tension
in the cable if the reactive force
on the upper pin has no horizontal
component?

Figure 3.14

SOLUTION. In the free-body diagram of the sign (Fig. 3.14), the
unknown reaction of the lower pin
is resolved into a horizontal
component F_1 and a vertical component F_2. The reaction of the
upper pin is the purely vertical
force F_3. The moment equation
about the lower pin is

$$- (200)(33) + (T \sin 35°)(69) + (T \cos 35°)(42) = 0$$

from which it follows that

$$T = \frac{(200)(33)}{69 \sin 35° + 42 \cos 35°} = 89 \text{ pounds.}$$

Remarks: 1. By taking moments about the lower pin, none of the
unknown forces F_1, F_2, F_3 appear in the moment
equation since their lines of action pass through
the lower pin.

2. Notice how Varignon's Theorem was used in computing
the moment of the force T of the cable by resolving
it into a horizontal force and a vertical force and
adding the individual moments.

EXERCISES

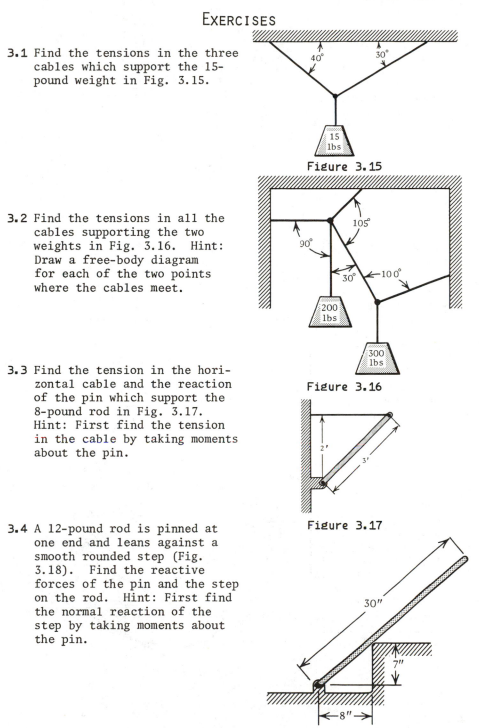

3.1 Find the tensions in the three cables which support the 15-pound weight in Fig. 3.15.

Figure 3.15

3.2 Find the tensions in all the cables supporting the two weights in Fig. 3.16. Hint: Draw a free-body diagram for each of the two points where the cables meet.

Figure 3.16

3.3 Find the tension in the horizontal cable and the reaction of the pin which support the 8-pound rod in Fig. 3.17. Hint: First find the tension in the cable by taking moments about the pin.

Figure 3.17

3.4 A 12-pound rod is pinned at one end and leans against a smooth rounded step (Fig. 3.18). Find the reactive forces of the pin and the step on the rod. Hint: First find the normal reaction of the step by taking moments about the pin.

Figure 3.18

3.5 Find the reactions of the three pins on the two weight-less rods which support the 50-pound weight in Fig. 3.19. Hint: Draw separate free-body diagrams of the two rods.

Figure 3.19

3.6 Two slender five-pound rods are pinned together and stand on a smooth surface. A horizontal cable connects them together at their top ends. Find the tension in the cable, the normal forces of the surface, and the reactive forces of the pin on each rod. Hint: By symmetry, the two normal forces of the surface are equal.

Figure 3.20

3.7 Show that two parallel forces F and -F of equal magnitude and opposite direction have moment Fd or $-Fd$ where d is the distance between their lines of action, and that the moment is independent of the point about which it is computed. (Such a pair of forces is called a *couple*.)

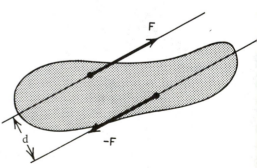

Figure 3.21

3.8 Show that if exactly two forces act on a rigid body in equilibrium, then they must be of equal magnitude and opposite direction and have the same line of action.

3.9 (a) Show that if exactly three forces act on a rigid body in equilibrium, then their lines of action must be either concurrent or parallel.

    (b) Use the result in part (a) to solve Example 3.2 without computing any moments.

    (c) Use the result in part (a) to solve Example 3.3 without computing any moments.

4

Graph Theory

*A matrix representation of relationships
existing between members of a set is introduced. Through simple matrix arithmetic, the
representation is used to analyze and clarify
these relationships.*

PREREQUISITES: Matrix multiplication and addition

INTRODUCTION

There are countless examples of sets consisting of a finite
number of members in which some kind of relation exists between members of the set. For example, the set could consist of a collection
of people, animals, countries, companies, sports teams or cities;
and the relation between two members, A and B, from such a set could
be that person A dominates person B, animal A feeds on animal B,
country A militarily supports country B, company A sells its product
to company B, sports team A consistantly beats sports team B, or
city A has a direct airline flight to city B.

In this chapter, we shall study the theory of directed graphs
to mathematically model the types of sets and relations in the above
examples.

DIRECTED GRAPHS

By a directed graph we mean a finite set of elements, P_1, P_2, ... , P_n, together with a finite collection of ordered pairs, (P_i, P_j), of distinct elements of this set, with no ordered pair being repeated. The elements of the set are called *vertices*, and the ordered pairs are called *directed edges*, of the directed graph. We use the notation $P_i \rightarrow P_j$ (read "P_i is connected to P_j") to indicate that the directed edge (P_i, P_j) belongs to the directed graph. Graphically, we can visualize a directed graph (Fig. 4.1) by representing the vertices as points in the plane, and representing $P_i \rightarrow P_j$ by drawing a line or arc from vertex P_i to vertex P_j, with an arrow pointing from P_i to P_j. If both $P_i \rightarrow P_j$ and $P_j \rightarrow P_i$ hold (denoted $P_i \leftrightarrow P_j$), we draw a single line between P_i and P_j with two oppositely pointing arrows (as with P_2 and P_3 in the figure).

Notice, as in Fig. 4.1, that a directed graph may have separate "components" of vertices which are connected only among themselves; and some vertices, such as P_5, may not be connected with any other vertex. Also, since $P_i \rightarrow P_i$ is not permitted in a directed graph, a vertex cannot be connected with itself by a single arc which does not pass through any other vertex.

In Fig. 4.2 we have drawn diagrams representing three more examples of directed graphs.

Figure 4.1

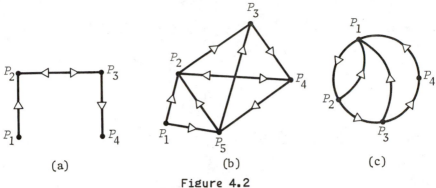

(a)   (b)   (c)

Figure 4.2

With a directed graph having n vertices, we may associate an $n \times n$ matrix $M = [m_{ij}]$, called the *vertex matrix* of the directed graph. Its elements are defined by

$$m_{ij} = \begin{cases} 1, & \text{if } P_i \rightarrow P_j \\ 0, & \text{otherwise} \end{cases}$$

for $i, j = 1, 2, \ldots n$. For the three directed graphs in Fig. 4.2, the corresponding vertex matrices are:

Figure 4.2(a):

$$M = \begin{bmatrix} 0 & 1 & 0 & 0 \\ 0 & 0 & 1 & 0 \\ 0 & 1 & 0 & 1 \\ 0 & 0 & 0 & 0 \end{bmatrix}$$

Figure 4.2(b):

$$M = \begin{bmatrix} 0 & 1 & 0 & 0 & 1 \\ 0 & 0 & 1 & 1 & 0 \\ 0 & 0 & 0 & 1 & 0 \\ 0 & 1 & 0 & 0 & 1 \\ 0 & 1 & 1 & 0 & 0 \end{bmatrix}$$

Figure 4.2(c):

$$M = \begin{bmatrix} 0 & 1 & 0 & 0 \\ 1 & 0 & 1 & 0 \\ 1 & 0 & 0 & 1 \\ 1 & 0 & 0 & 0 \end{bmatrix}$$

By their definition, vertex matrices have the following two properties:

(i)   All entries are either 0 or 1.
(ii) All diagonal entries are 0.

Conversely, any matrix with these two properties determines a unique directed graph having the matrix as its vertex matrix. For example, the matrix

$$M = \begin{bmatrix} 0 & 1 & 1 & 0 \\ 0 & 0 & 1 & 0 \\ 1 & 0 & 0 & 1 \\ 0 & 0 & 0 & 0 \end{bmatrix}$$

determines the directed graph in Fig. 4.3.

Figure 4.3

EXAMPLE 4.1

A certain family consists of a mother, father, daughter, and two sons. The family members have influence, or power, over each other in the following ways: the mother can influence the daughter and the oldest son; the father can influence the two sons; the daughter can influence the father; the oldest son can influence the youngest son; and the youngest son can influence the mother. We may model this family influence pattern with a directed graph whose vertices are the five family members. If family member A influences family member B, we write $A \rightarrow B$. Figure 4.4 is the resulting directed graph, where we have used obvious letter designations for the five family members. The vertex matrix of this directed graph is

	M	F	D	OS	YS
M	0	0	1	1	0
F	0	0	0	1	1
D	0	1	0	0	0
OS	0	0	0	0	1
YS	1	0	0	0	0

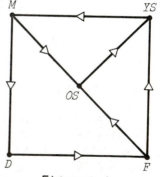

Figure 4.4

EXAMPLE 4.2 In chess, the knight
moves in an "L"-shaped pattern
about the chessboard. It may move
horizontally two squares and then
vertically one square, or it may
move vertically two squares and
then horizontally one square. For
example, from the black square in
Fig. 4.5, the knight may move to
any of the eight shaded squares.
Suppose the knight is restricted
to the nine numbered squares in
Fig. 4.6. If by $i \rightarrow j$ we mean that
the knight may move from square i
to square j, the directed graph in
Figure 4.7 illustrates all possible moves that the knight may make
among these nine squares. In Fig.
4.8 we have "unraveled" Fig. 4.7
to make the pattern of possible
moves clearer.

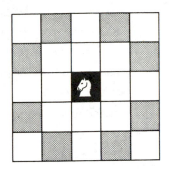

Figure 4.5

1	2	3
4	5	6
7	8	9

Figure 4.6

The vertex matrix of this
directed graph is given by

$$M = \begin{bmatrix} 0 & 0 & 0 & 0 & 0 & 1 & 0 & 1 & 0 \\ 0 & 0 & 0 & 0 & 0 & 0 & 1 & 0 & 1 \\ 0 & 0 & 0 & 1 & 0 & 0 & 0 & 1 & 0 \\ 0 & 0 & 1 & 0 & 0 & 0 & 0 & 0 & 1 \\ 0 & 0 & 0 & 0 & 0 & 0 & 0 & 0 & 0 \\ 1 & 0 & 0 & 0 & 0 & 0 & 1 & 0 & 0 \\ 0 & 1 & 0 & 0 & 0 & 1 & 0 & 0 & 0 \\ 1 & 0 & 1 & 0 & 0 & 0 & 0 & 0 & 0 \\ 0 & 1 & 0 & 1 & 0 & 0 & 0 & 0 & 0 \end{bmatrix}.$$

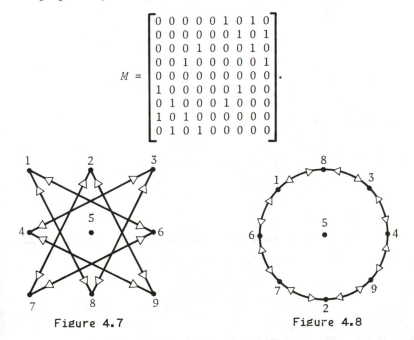

Figure 4.7

Figure 4.8

In Example 4.1, the father cannot directly influence the mother; i.e., $F \rightarrow M$ is not true. But he can influence the youngest son, who can then influence the mother. We write this as $F \rightarrow YS \rightarrow M$, and call it a *2-step connection* from F to M. Analogously, we call $M \rightarrow D$ a *1-step connection*; $F \rightarrow OS \rightarrow YS \rightarrow M$ a *3-step connection*; and so forth. Let us now consider a technique for finding the number of all possible r-step connections ($r = 1, 2, \ldots$) from one vertex P_i to another vertex P_j of an arbitrary directed graph. (This will include the case when P_i and P_j are the same vertex.) The number of 1-step connections from P_i to P_j is simply m_{ij}. That is, there is either zero or one 1-step connection from P_i to P_j depending on whether m_{ij} is zero or one. For the number of 2-step connections, we consider the square of the vertex matrix. If we let $m_{ij}^{(2)}$ be the (i, j)-th element of M^2, we have

$$m_{ij}^{(2)} = m_{i1}m_{1j} + m_{i2}m_{2j} + \cdots + m_{in}m_{nj}. \qquad (4.1)$$

Now, if $m_{i1} = m_{1j} = 1$, there is a 2-step connection $P_i \rightarrow P_1 \rightarrow P_j$ from P_i to P_j. But if either m_{i1} or m_{1j} is zero, such a 2-step connection is not possible. Thus $P_i \rightarrow P_1 \rightarrow P_j$ is a 2-step connection if and only if $m_{i1}m_{1j} = 1$. Similarly, for any $k = 1, 2, \ldots, n$, $P_i \rightarrow P_k \rightarrow P_j$ is a 2-step connection from P_i to P_j if and only if the term $m_{ik}m_{kj}$ on the righthand side of (4.1) is one; otherwise the term is zero. Thus, the righthand side of (4.1) is the total number of two 2-step connections from P_i to P_j.

This same argument will work for finding the number of 3-, 4-,.... step connections from P_i to P_j. The general result is the following:

THEOREM 4.1 *Let M be the vertex matrix of a directed graph and let $m_{ij}^{(r)}$ be the (i, j)-th element of M^r. Then $m_{ij}^{(r)}$ is equal to the number of r-step connections from P_i to P_j.*

EXAMPLE 4.3 Figure 4.9 is the route map of a small airline which services the four cities P_1, P_2, P_3, and P_4. As a directed graph, its vertex matrix is

$$M = \begin{bmatrix} 0 & 1 & 1 & 0 \\ 1 & 0 & 1 & 0 \\ 1 & 0 & 0 & 1 \\ 0 & 1 & 1 & 0 \end{bmatrix}.$$

We have that

$$M^2 = \begin{bmatrix} 2 & 0 & 1 & 1 \\ 1 & 1 & 1 & 1 \\ 0 & 2 & 2 & 0 \\ 2 & 0 & 1 & 1 \end{bmatrix}$$

and

$$M^3 = \begin{bmatrix} 1 & 3 & 3 & 1 \\ 2 & 2 & 3 & 1 \\ 4 & 0 & 2 & 2 \\ 1 & 3 & 3 & 1 \end{bmatrix}.$$

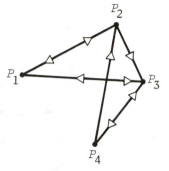

Figure 4.9

If we are interested in connections from city P_4 to city P_3, we may use Theorem 4.1 to find their number. Since $m_{43} = 1$, there is one 1-step connection; since $m_{43}^{(2)} = 1$, there is one 2-step connection; and since $m_{43}^{(3)} = 3$, there are three 3-step connections. To verify this, from Fig. 4.9 we find:

1-step connections from P_4 to P_3: $P_4 \to P_3$

2-step connections from P_4 to P_3: $P_4 \to P_2 \to P_3$

3-step connections from P_4 to P_3: $P_4 \to P_3 \to P_4 \to P_3$
$P_4 \to P_2 \to P_1 \to P_3$
$P_4 \to P_3 \to P_1 \to P_3$.

CLIQUES

Let us introduce the following definition:

> **DEFINITION 4.1** *A subset of a directed graph is called a*
> **clique** *if it satisfies the following three conditions:*
> (i)   *The subset contains at least three vertices.*
> (ii) *For each pair of vertices P_i and P_j in the sub-*
> *set, both $P_i \rightarrow P_j$ and $P_j \rightarrow P_i$ are true.*
> (iii) *The subset is as large as possible; i.e., it is*
> *not possible to add another vertex to the subset*
> *and still satisfy condition (ii).*

This definition suggests that cliques are maximal subsets which are
in perfect "communication" with each other. For example, if the
vertices represent cities, and $P_i \rightarrow P_j$ means that there is a direct
airline flight from city P_i to city P_j, then there is a direct
flight between any two cities within a clique in either direction.

EXAMPLE 4.4 The directed graph
illustrated in Fig. 4.10 (which
might represent the route map of
an airline) has two cliques:

$\{P_1,\ P_2,\ P_3,\ P_4\}$ and $\{P_3,\ P_4,\ P_6\}$.

This example shows that a directed
graph may contain several cliques,
and that a vertex may simultaneously belong to more than one
clique. ◆━◆━◆

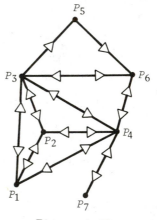

Figure 4.10

For simple directed graphs,
cliques may be found by inspection.
But for large directed graphs, it
would be desirable to have a systematic procedure for detecting
cliques. We shall now discuss a theorem which identifies those vertices which belong to cliques. First, we define a matrix $S = [s_{ij}]$
related to a given directed graph as follows:

$$s_{ij} = \begin{cases} 1, & \text{if } P_i \leftrightarrow P_j \\ 0, & \text{otherwise}. \end{cases}$$

The matrix S determines a directed graph which is the same as the given directed graph with the exception that the directed edges with only one arrow are deleted. For example, if the original directed graph is given by Fig. 4.11(a),

(a)   (b)

Figure 4.11

the directed graph which has S as its vertex matrix is given in Fig. 4.11(b). Alternately, S may be obtained from the vertex matrix M of the original directed graph by setting $s_{ij} = 1$ if $m_{ij} = m_{ji} = 1$, and otherwise setting $s_{ij} = 0$.

Using this matrix S, the theorem referred to is

THEOREM 4.2 *Let $s_{ij}^{(3)}$ be the (i, j)-th element of S^3. Then a vertex P_i belongs to some clique if and only if $s_{ii}^{(3)} \neq 0$.*

PROOF If $s_{ii}^{(3)} \neq 0$, then there is at least one 3-step connection from P_i to itself in the modified directed graph determined by S. Suppose it is $P_i \rightarrow P_j \rightarrow P_k \rightarrow P_i$. In the modified directed graph, all directed relations are two-way, so that we also have $P_i \leftrightarrow P_j \leftrightarrow P_k \leftrightarrow P_i$. But this means either $\{P_i, P_j, P_k\}$ is a clique, or is a subset of a clique. In either case, P_i must belong to some clique. The converse statement that if P_i belongs to a clique then $s_{ii}^{(3)} \neq 0$ follows in a similar manner. □

EXAMPLE 4.5 Suppose a directed graph has as its vertex matrix

$$M = \begin{bmatrix} 0 & 1 & 1 & 1 \\ 1 & 0 & 1 & 0 \\ 0 & 1 & 0 & 1 \\ 1 & 0 & 0 & 0 \end{bmatrix}.$$

Then

$$S = \begin{bmatrix} 0 & 1 & 0 & 1 \\ 1 & 0 & 1 & 0 \\ 0 & 1 & 0 & 0 \\ 1 & 0 & 0 & 0 \end{bmatrix} \quad \text{and} \quad S^3 = \begin{bmatrix} 0 & 3 & 0 & 2 \\ 3 & 0 & 2 & 0 \\ 0 & 2 & 0 & 1 \\ 2 & 0 & 1 & 0 \end{bmatrix}.$$

Since all diagonal entries of S^3 are zero, it follows from Theorem 4.2 that the directed graph has no cliques.

EXAMPLE 4.6 Suppose a directed graph has as its vertex matrix

$$M = \begin{bmatrix} 0 & 1 & 0 & 1 & 1 \\ 1 & 0 & 0 & 1 & 0 \\ 1 & 1 & 0 & 1 & 0 \\ 1 & 1 & 0 & 0 & 0 \\ 1 & 0 & 0 & 1 & 0 \end{bmatrix}.$$

Then

$$S = \begin{bmatrix} 0 & 1 & 0 & 1 & 1 \\ 1 & 0 & 0 & 1 & 0 \\ 0 & 0 & 0 & 0 & 0 \\ 1 & 1 & 0 & 0 & 0 \\ 1 & 0 & 0 & 0 & 0 \end{bmatrix} \quad \text{and} \quad S^3 = \begin{bmatrix} 2 & 4 & 0 & 4 & 3 \\ 4 & 2 & 0 & 3 & 1 \\ 0 & 0 & 0 & 0 & 0 \\ 4 & 3 & 0 & 2 & 1 \\ 3 & 1 & 0 & 1 & 0 \end{bmatrix}.$$

The nonzero diagonal entries of S^3 are $s_{11}^{(3)}$, $s_{22}^{(3)}$, and $s_{44}^{(3)}$. Consequently, in the given directed graph, P_1, P_2, and P_4 belong to cliques. Since a clique must contain at least three vertices, the directed graph has only one clique; namely, $\{P_1, P_2, P_4\}$.

DOMINANCE DIRECTED GRAPHS

In many groups of individuals or animals, there is a definite "pecking order" or dominance relation between any two members of the group. That is, given any two individuals A and B, either A dominates B or B dominates A, but not both. In terms of a directed graph in which $P_i \rightarrow P_j$ means P_i dominates P_j, this means that for all distinct pairs, either $P_i \rightarrow P_j$ or $P_j \rightarrow P_i$, but not both. In general, we have the following definition:

DEFINITION 4.2 *A* **dominance directed graph** *is a directed graph such that for any distinct pair of vertices P_i and P_j, either $P_i \rightarrow P_j$ or $P_j \rightarrow P_i$, but not both.*

An example which provides a directed graph satisfying this definition is a league of n sports teams which play each other exactly one time, as in one round of a round-robin tournament in which no ties are allowed. If $P_i \rightarrow P_j$ means that team P_i beat team P_j in their single match, it is easily seen that Definition 4.2 is satisfied. For this reason, dominance directed graphs are sometimes called *tournaments*.

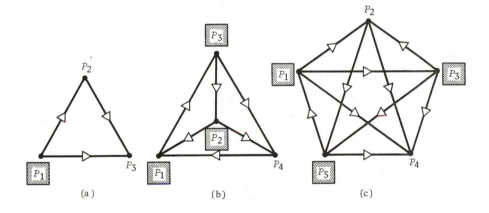

Figure 4.12

In Fig. 4.12 there are illustrated some dominance directed graphs with three, four, and five vertices, respectively. In these three graphs, the shaded vertices have the following interesting property: from each one there is either a 1-step or a 2-step connection to any other vertex in its graph. In a sense, these vertices are more "powerful" than the vertices which do not have this property. In a sports tournament, such a vertex corresponds to a team A such that if B is any other team, either A beat B, or beat a team which beat B. We shall now state and prove a theorem that guarantees that any dominance directed graph has at least one vertex with this property.

THEOREM 4.3 *In any dominance directed graph there is at least one vertex from which there is a 1-step or 2-step connection to any other vertex.*

PROOF. Consider a vertex (there may be several) with the largest total number of 1-step and 2-step connections to other vertices in the graph. By renumbering the vertices, we may assume that P_1 is such a vertex. Suppose there is some vertex P_i such that there is no 1-step or 2-step connection from P_1 to P_i. Then, in particular, $P_1 \to P_i$ is not true, so that, by Definition 4.2, it must be that $P_i \to P_1$. Next, let P_k be any vertex such that $P_1 \to P_k$ is true. Then we cannot have $P_k \to P_i$, since then $P_1 \to P_k \to P_i$ would be a 2-step connection from P_1 to P_i. Thus, it must be that $P_i \to P_k$. That is, P_i has 1-step connections to all the vertices to which P_1 has 1-step connections. The vertex P_i must then also have 2-step connections to all the vertices to which P_1 has 2-step connections. But since, in addition, we have that $P_i \to P_1$, this means that P_i has more 1-step and 2-step connections to other vertices than does P_1. However, this is a contradiction to the way P_1 was chosen. Hence, there can be no vertex P_i to which P_1 has no 1-step or 2-step connection.□

This proof shows that a vertex with the largest total number of 1-step and 2-step connections to other vertices has the property stated in the theorem. There is a simple way of finding such vertices by using the vertex matrix M and its square M^2. The sum of the entries in the i-th row of M is the total number of 1-step connections from P_i to other vertices, and the sum of the entries of the i-th row of M^2 is the total number of 2-step connections from P_i to other vertices. Consequently, the sum of the entries of the i-th row of the matrix $A = M + M^2$ is the total number of 1-step and 2-step connections from P_i to other vertices. In other words, a row of $A = M + M^2$ with the largest row sum identifies a vertex having the property stated in Theorem 4.3.

EXAMPLE 4.7 Suppose five baseball teams play each other exactly
once, and the results are as indicated in the dominance directed
graph of Fig. 4.13. The vertex
matrix of the graph is

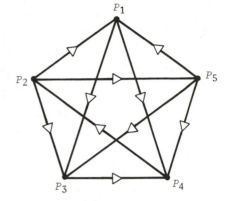

P_1

P_2 P_5

P_3 P_4

$$M = \begin{bmatrix} 0 & 0 & 1 & 1 & 0 \\ 1 & 0 & 1 & 0 & 1 \\ 0 & 0 & 0 & 1 & 0 \\ 0 & 1 & 0 & 0 & 0 \\ 1 & 0 & 1 & 1 & 0 \end{bmatrix}.$$

And so

Figure 4.13

$$A = M + M^2 = \begin{bmatrix} 0 & 0 & 1 & 1 & 0 \\ 1 & 0 & 1 & 0 & 1 \\ 0 & 0 & 0 & 1 & 0 \\ 0 & 1 & 0 & 0 & 0 \\ 1 & 0 & 1 & 1 & 0 \end{bmatrix} + \begin{bmatrix} 0 & 1 & 0 & 1 & 0 \\ 1 & 0 & 2 & 3 & 0 \\ 0 & 1 & 0 & 0 & 0 \\ 1 & 0 & 1 & 0 & 1 \\ 0 & 1 & 1 & 2 & 0 \end{bmatrix} = \begin{bmatrix} 0 & 1 & 1 & 2 & 0 \\ 2 & 0 & 3 & 3 & 1 \\ 0 & 1 & 0 & 1 & 0 \\ 1 & 1 & 1 & 0 & 1 \\ 1 & 1 & 2 & 3 & 0 \end{bmatrix}.$$

The row sums of A are:

1st row sum = 4
2nd row sum = 9
3rd row sum = 2
4th row sum = 4
5th row sum = 7 .

Since the second row has the largest row sum, the vertex P_2 must
have a 1-step or 2-step connection to any other vertex. This is
easily verified from Fig. 4.13.

We have informally suggested that a vertex with the largest
number of 1-step and 2-step connections to other vertices is a "powerful" vertex. Let us formalize this concept with the following
definition:

DEFINITION 4.3 *The* **power** *of a vertex of a dominance directed graph is the total number of 1-step and 2-step
connections from it to other vertices. Alternately, the
power of a vertex P_i is the sum of the entries of the
i-th row of the matrix $A = M + M^2$, where M is the vertex
matrix of the directed graph.*

EXAMPLE 4.8 Let us rank the five baseball teams in Example 4.7 according to their powers. From the calculations for the row sums in that example we have:

$$\text{Power of team } P_1 = 4$$
$$\text{Power of team } P_2 = 9$$
$$\text{Power of team } P_3 = 2$$
$$\text{Power of team } P_4 = 4$$
$$\text{Power of team } P_5 = 7$$

Hence, the ranking of the teams according to their powers would be: first, P_2; second, P_5; third, P_1 and P_4 (tie); fifth, P_3.

EXERCISES

4.1 Construct the vertex matrix for each of the directed graphs illustrated in Fig. 4.14.

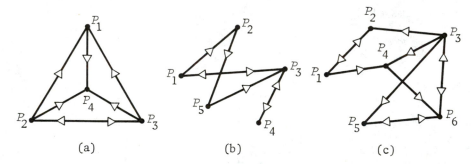

(a) (b) (c)

Figure 4.14

4.2 Draw a diagram of the directed graph corresponding to each of the following vertex matrices.

(a) $\begin{bmatrix} 0 & 1 & 1 & 0 \\ 1 & 0 & 0 & 0 \\ 0 & 0 & 0 & 1 \\ 1 & 0 & 1 & 0 \end{bmatrix}$
(b) $\begin{bmatrix} 0 & 0 & 1 & 0 & 0 \\ 1 & 0 & 0 & 0 & 1 \\ 0 & 1 & 0 & 1 & 1 \\ 0 & 0 & 0 & 0 & 0 \\ 1 & 1 & 1 & 0 & 0 \end{bmatrix}$
(c) $\begin{bmatrix} 0 & 1 & 0 & 1 & 0 & 1 \\ 1 & 0 & 0 & 0 & 1 & 0 \\ 0 & 0 & 0 & 0 & 0 & 0 \\ 1 & 1 & 0 & 0 & 1 & 0 \\ 0 & 0 & 0 & 1 & 0 & 1 \\ 0 & 1 & 0 & 0 & 1 & 0 \end{bmatrix}$

4.3 Let M be the following vertex matrix of a directed graph:

$$\begin{bmatrix} 0 & 1 & 1 & 1 \\ 1 & 0 & 0 & 0 \\ 0 & 1 & 0 & 1 \\ 0 & 1 & 1 & 0 \end{bmatrix}.$$

(a) Draw a diagram of the directed graph.
(b) Use Theorem 4.1 to find the number of 1-, 2-, and 3-step connections from the vertex P_1 to the vertex P_2. Verify your answer by listing the various connections as in Example 4.3.
(c) Repeat part (b) for the 1-, 2-, and 3-step connections from P_1 to P_4.

4.4 By inspection, locate all cliques in each of the directed graphs illustrated in Fig. 4.15.

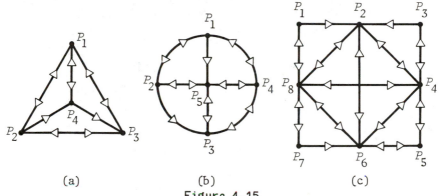

(a)　　　　　　　(b)　　　　　　　(c)

Figure 4.15

2.5 For each of the following vertex matrices, use Theorem 4.2 to find any cliques in the corresponding directed graphs.

(a) $\begin{bmatrix} 0 & 1 & 0 & 1 & 0 \\ 1 & 0 & 1 & 0 & 1 \\ 0 & 1 & 0 & 1 & 1 \\ 1 & 0 & 0 & 0 & 1 \\ 1 & 0 & 1 & 1 & 0 \end{bmatrix}$ (b) $\begin{bmatrix} 0 & 1 & 0 & 1 & 1 & 0 \\ 1 & 0 & 1 & 0 & 1 & 1 \\ 0 & 1 & 0 & 1 & 0 & 1 \\ 1 & 0 & 1 & 0 & 1 & 1 \\ 0 & 1 & 0 & 1 & 0 & 0 \\ 0 & 0 & 1 & 1 & 1 & 0 \end{bmatrix}$

4.6 For the dominance directed
graph illustrated in Fig.
2.16, construct the vertex
matrix and use Definition
2.3 to find the power of
each vertex.

4.7 Five baseball teams play each
other one time with the following results:

Figure 4.16

A beats B,C,D
B beats C,E
C beats D,E
D beats B
E beats A,D.

Rank the five baseball teams in accordance with the powers of
the vertices they correspond to in the dominance directed graph
representing the outcomes of the games.

Theory of Games

A general game in which two competing players choose separate strategies to reach opposing objectives is discussed. Using matrix techniques, the optimal strategy of each player is found for a few special cases.

PREREQUISITES: Matrix multiplication

INTRODUCTION

To introduce the basic concepts in the theory of games, let us consider the following carnival-type game which two people agree to play. We shall call the participants of the game player R and player C. Each player has a stationary wheel with a movable pointer on it as in Fig. 5.1. For reasons which will become clear, we shall call player R's wheel the row-wheel, and player C's wheel the column-wheel. The row-wheel is divided into three sectors numbered 1, 2, and 3, and the column-wheel is divided into four sectors numbered 1, 2, 3, and 4. The fractions of the area occupied by the various sectors are indicated in the figure. To play the game, each player spins the pointer of his wheel and lets it come to rest at random. The number of the sector in which each pointer comes to

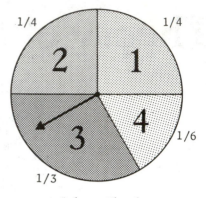

Row-Wheel
of Player R

Column-Wheel
of Player C

Figure 5.1

rest is called the *move* of that player. Thus, player R has three
possible moves, and player C has four possible moves. Depending on
the move each player makes, player C then makes a payment of money
to player R according to the following table:

		Player C's move			
		1	2	3	4
Player R's move	1	$3	$5	-$2	-$1
	2	-$2	$4	-$3	-$4
	3	$6	-$5	$0	$3

Table 5.1

For example, if the row-wheel pointer comes to rest in sector 1
(player R makes move 1), and the column-wheel pointer comes to rest
in sector 2 (player C makes move 2), then player C must pay player
R the sum of $5. Some of the entries in this table are negative,
indicating that player C makes a negative payment to player R. By
this we mean that player R makes a positive payment to player C.
For example, if the row-wheel shows 2 and the column wheel shows 4,
then player R pays player C the sum of $4, since the corresponding
entry in the table is -$4. In this way, the positive entries of the
table are the gains of player R and the losses of player C, and the
negative entries are the gains of player C and the losses of player
R.

In this game, the players have no control over their moves; each move is determined by chance. However, if each player can decide whether or not he wants to play, then each would want to know how much he can expect to win or lose over the long term if he chooses to play. We discuss this question in the first part of the next section. Later in the section we consider a more complicated situation in which the players can exercise some control over their moves by varying the sectors of their wheels.

Two-person Zero-sum Matrix Games

The game described above is an example of a two-person zero-sum matrix game. The term "zero-sum" means that in each play of the game the positive gain of one player is equal to the negative gain (loss) of the other player. That is, the sum of the two algebraic gains is zero. The term "matrix game" is used to describe a two-person game in which each player has only a finite number of moves, so that all possible outcomes of each play, and the corresponding gains of the players, may be displayed in tabular or matrix form, as in Table 5.1.

In a general game of this type, let player R have m possible moves and let player C have n possible moves. In a play of the game, each player makes one of his possible moves, and then a *payoff* is made from player C to player R depending on the moves. For $i = 1, 2, \ldots, m$ and $j = 1, 2, \ldots, n$, let us set

a_{ij} = payoff that player C makes to player R if player R makes move i and player C makes move j.

This payoff need not be money; it may be any type of commodity to which we can attach a numerical value. As before, if an entry a_{ij} is negative, we mean that player C receives a payoff of $|a_{ij}|$ from player R. We arrange these mn possible payoffs in the form of an $m \times n$ matrix

$$A = \begin{bmatrix} a_{11} & a_{12} & \cdots & a_{1n} \\ a_{21} & a_{22} & \cdots & a_{2n} \\ \vdots & \vdots & & \vdots \\ a_{m1} & a_{m2} & \cdots & a_{mn} \end{bmatrix}$$

which we shall call the *payoff matrix* of the game.

Each player is to make his moves on a probabilistic basis. For example, for the game discussed in the introduction, the proportion of the area of a sector to the area of the wheel would be the probability that the player makes the move corresponding to that sector.

Thus, from Fig. 5.1, we see that player R would make move 2 with probability 1/3, and player C would make move 2 with probability 1/4. In the general case, let us set

p_i = probability that player R makes move i $(i = 1, 2, \ldots, m)$,

and

q_j = probability that player C makes move j $(j = 1, 2, \ldots, n)$.

By their definitions, we have that

$$p_1 + p_2 + \cdots + p_m = 1$$

and

$$q_1 + q_2 + \cdots + q_n = 1.$$

With these probabilities, p_i and q_j, we form two vectors

$$\mathbf{p} = \begin{bmatrix} p_1 & p_2 & \cdots & p_m \end{bmatrix} \quad \text{and} \quad \mathbf{q} = \begin{bmatrix} q_1 \\ q_2 \\ \vdots \\ q_n \end{bmatrix}.$$

We call the row vector $\mathbf{p}$ the *strategy of player R*, and call the column vector $\mathbf{q}$ the *strategy of player C*. For example, from Fig. 5.1 we have

$$\mathbf{p} = \begin{bmatrix} 1/6 & 1/3 & 1/2 \end{bmatrix} \quad \text{and} \quad \mathbf{q} = \begin{bmatrix} 1/4 \\ 1/4 \\ 1/3 \\ 1/6 \end{bmatrix}$$

for the game described in the introduction.

From the theory of probability, if the probability that player R makes move i is p_i, and independently the probability that player C makes move j is q_j, then $p_i q_j$ is the probability that for any one play of the game player R makes move i and player C makes move j. The payoff to player R for such a pair of moves is a_{ij}. Thus $p_i q_j$ is also the probability that for any one play of the game the payoff to player R is entry a_{ij}. If we multiply each possible payoff by its corresponding probability and sum over all possible payoffs, we obtain the expression

$$a_{11}p_1q_1 + a_{12}p_1q_2 + a_{21}p_2q_1 + \cdots + a_{mn}p_mq_n. \tag{5.1}$$

Equation (5.1) is a weighted average of the payoffs to player R, with each payoff being weighted according to the probability of its occurrence. In the theory of probability this weighted average is called the *expected payoff* to player R. It can be shown that if the game is played many times, the long term average payoff per play to player R is given by this expression. We shall denote this expected payoff by $E(\mathbf{p}, \mathbf{q})$ to emphasize the fact that it depends on the strategies of the two players. From the definition of the payoff matrix A and the strategies $\mathbf{p}$ and $\mathbf{q}$, it may be verified that we may express the expected payoff in matrix notation as

$$E(\mathbf{p}, \mathbf{q}) = \begin{bmatrix} p_1 & p_2 & \cdots & p_m \end{bmatrix} \begin{bmatrix} a_{11} & a_{12} & \cdots & a_{1n} \\ a_{21} & a_{22} & \cdots & a_{2n} \\ \vdots & \vdots & & \vdots \\ a_{m1} & a_{m2} & \cdots & a_{mn} \end{bmatrix} \begin{bmatrix} q_1 \\ q_2 \\ \vdots \\ q_n \end{bmatrix} = \mathbf{p}A\mathbf{q}. \tag{5.2}$$

Observe that since $E(\mathbf{p}, \mathbf{q})$ is the expected payoff to player R, then $-E(\mathbf{p}, \mathbf{q})$ is the expected payoff to player C.

EXAMPLE 5.1 For the game described in the introduction, we have

$$E(\mathbf{p}, \mathbf{q}) = \mathbf{p}A\mathbf{q} = \begin{bmatrix} 1/6 & 1/3 & 1/2 \end{bmatrix} \begin{bmatrix} 3 & 5 & -2 & -1 \\ -2 & 4 & -3 & -4 \\ 6 & -5 & 0 & 3 \end{bmatrix} \begin{bmatrix} 1/4 \\ 1/4 \\ 1/3 \\ 1/6 \end{bmatrix} = 13/72 = .1806\ldots$$

Thus, in the long run player R can expect to receive an average of about 18¢ from player C in each play of the game.

So far we have been discussing the situation in which each player has a predetermined strategy. We shall now consider the more difficult situation in which both players may change their strategies independently. For example, in the game described in the introduction, we would allow both players to alter the areas of the sectors of their wheels and thereby control the probabilities of their respective moves. This qualitatively changes the nature of the problem and puts us firmly into the field of true game theory. It is understood that neither player knows the strategy the other will choose. It will also be assumed that each player will make the best possible choice of strategy, and that the other player knows this. Thus player R attempts to choose a strategy p such that $E(p, q)$ is as large as possible for the best strategy q that player C can choose; and similarly player C attempts to choose a strategy q such that $E(p, q)$ is as small as possible for the best strategy p that player R can choose. First, to see that such choices are possible, let us state the following theorem, called the *Fundamental Theorem of 2-person Zero-sum Games*:

THEOREM 5.1 *There exist strategies* p* *and* q* *such that*

$$E(p^*, q) \geq E(p^*, q^*) \geq E(p, q^*) \tag{5.3}$$

for all strategies p *and* q.

(The proof of this theorem involves ideas from the theory of Linear Programming, and we shall not give it here. In the next two sections we shall prove the theorem for two special cases.)

This result guarantees that each player has a best possible strategy. To see why, let p* and q* be the strategies stated in the theorem, and let $v = E(p^*, q^*)$. The lefthand inequality of Eq. (5.3) then reads

$$E(p^*, q) \geq v, \qquad \text{for all strategies } q.$$

This means that if player R chooses the strategy p*, then no matter what strategy q player C chooses, the expected payoff to player R will never be below v. Furthermore, the expected payoff v is the best possible. For suppose there is some strategy p** that player R can choose such that

$$E(p^{**}, q) > v, \qquad \text{for all strategies } q.$$

Then, in particular,

$$E(p^{**}, q^*) > v.$$

But this is a contradiction to the righthand inequality of Eq. (5.3) which requires that $E(p^{**}, q^*) \leq v$. Consequently, the best player R can do is prevent his expected payoff from falling below the value v.

Similarly, if player C chooses the strategy q*, then from the righthand inequality of Eq. (5.3) we have

$$v \geq E(p, q^*), \qquad \text{for all strategies } p.$$

That is, the expected payoff to player R never exceeds the value v if player C chooses the stragegy q^*, no matter what strategy player R chooses. This bound similarly cannot be lowered, since player R may always choose the strategy p^* to force his expected payoff to be at least the value v.

On the basis of this discussion, we make the following definitions:

DEFINITION 5.1 *If* p^* *and* q^* *are strategies such that*

$$E(p^*, q) \geq E(p^*, q^*) \geq E(p, q^*) \tag{5.4}$$

for all strategies p *and* q, *then*

(i) p^* *is called an* **optimal strategy for player R,**

(ii) q^* *is called an* **optimal strategy for player C,**

(iii) $v = E(p^*, q^*)$ *is called the* **value** *of the game.*

The wording in this definition suggests that optimal strategies are not necessarily unique. This is indeed the case, and in Exercise 5.2 we ask the reader to show this. However, it can be proved that any two sets of optimal strategies always result in the same value v of the game. That is, if p^*, q^* and p^{**}, q^{**} are optimal strategies, then

$$E(p^*, q^*) = E(p^{**}, q^{**}). \tag{5.5}$$

The value of a game is thus the expected payoff to player R when both players choose any possible optimal strategies.

To find optimal strategies, we must find vectors p^* and q^* which satisfy condition (5.4). This is generally done by using Linear Programming techniques. (See Chapter 20 for an introduction to this field.) In the next two sections we discuss special cases for which optimal strategies may be found by more elementary techniques.

STRICTLY DETERMINED GAMES

Let us introduce the following definition:

DEFINITION 5.2 *An entry* a_{rs} *in a payoff matrix A is called a* **saddle point** *if*

> (i) a_{rs} is the smallest entry in its row, and
> (ii) a_{rs} is the largest entry in its column.
> A game whose payoff matrix has a saddle point is called
> **strictly determined.**

For example, the boxed element in each of the following payoff matrices is a saddle point:

$$\begin{bmatrix} 3 & \boxed{1} \\ -4 & 0 \end{bmatrix} \qquad \begin{bmatrix} 30 & -50 & -5 \\ \boxed{60} & 90 & 75 \\ -10 & 60 & -30 \end{bmatrix} \qquad \begin{bmatrix} 0 & -3 & 5 & -9 \\ 15 & -8 & -2 & 10 \\ 7 & 10 & \boxed{6} & 9 \\ 6 & 11 & -3 & 2 \end{bmatrix}$$

If a matrix has a saddle point a_{rs}, it turns out that the following strategies are optimal strategies for the two players:

$$p^* = \begin{bmatrix} 0 & 0 & \cdots & 1 & \cdots & 0 \end{bmatrix} , \qquad q^* = \begin{bmatrix} 0 \\ 0 \\ \vdots \\ 1 \\ \vdots \\ 0 \end{bmatrix} \longleftarrow s\text{-th entry.}$$

$\nearrow$
r-th entry

That is, an optimal strategy for player R is to always make the r-th move, and an optimal strategy for player C is to always make the s-th move. Such strategies for which only one move is possible are called *pure strategies*. Strategies for which more than one move is possible are called *mixed strategies*. To show that the above pure strategies are optimal, the reader may verify the following three equations (see Exercise 5.6):

$$E(p^*, q^*) = p^*Aq^* = a_{rs} , \tag{5.6}$$

$$E(p^*, q) = p^*Aq \geq a_{rs} \text{ for any strategy } q, \tag{5.7}$$

$$E(p, q^*) = pAq^* \leq a_{rs} \text{ for any strategy } p. \tag{5.8}$$

Together, these three equations imply that

$$E(p^*, q) \geq E(p^*, q^*) \geq E(p, q^*)$$

for all strategies p and q. Since this is exactly condition (5.4), it follows that p^* and q^* are optimal strategies.

From Eq. (5.6), the value of a strictly determined game is simply the numerical value of a saddle point a_{rs}. It is possible for a payoff matrix to have several saddle points, but then the uniqueness of the value of a game guarantees that the numerical values of all saddle points are the same.

EXAMPLE 5.2

Two competing television networks, R and C, are scheduling 1-hour programs in the same time period. Network R can schedule one of three possible programs, and network C can schedule one of four possible programs. Neither network knows which program the other will schedule. Both networks ask the same outside polling agency to give them an estimate of how all possible pairings of the programs will divide the viewing audience. The agency gives them each the following table, whose (i,j)th entry is the percentage of the viewing audience that will watch network R if network R's program i is paired against network C's program j.

		Network C's Program			
		1	2	3	4
Network R's Program	1	60	20	30	55
	2	50	75	45	60
	3	70	45	35	30

What program should each network schedule in order to maximize its viewing audience?

SOLUTION

Let us subtract 50 from each entry in the above table to construct the following matrix:

$$\begin{bmatrix} 10 & -30 & -20 & 5 \\ 0 & 25 & -5 & 10 \\ 20 & -5 & -15 & -20 \end{bmatrix}.$$

This is the payoff matrix of the two-person zero-sum game in which each network is considered to start with 50% of the audience, and the (i,j)th entry of the matrix is the percentage of the viewing audience that network C loses to network R if programs i and j are paired against each other. It is easily seen that the entry

$$a_{23} = -5$$

is a saddle point of the payoff matrix. Hence, the optimal strategy of network R is to schedule program 2 and the optimal strategy of network C is to schedule program 3. This will result in network R receiving 45% of the audience and network C receiving 55% of the audience.

2 x 2 Matrix Games

Another case in which the optimal strategies may be found by elementary means is when each player has only two possible moves. In this case, the payoff matrix is a 2×2 matrix

$$A = \begin{bmatrix} a_{11} & a_{12} \\ a_{21} & a_{22} \end{bmatrix}.$$

If the game is strictly determined, at least one of the four entries of A is a saddle point, and the technique of the previous section can then be applied to determine optimal strategies for the two players. If the game is not strictly determined, we proceed as described below. First, we compute the expected payoff for arbitrary strategies $\mathbf{p}$ and $\mathbf{q}$:

$$E(\mathbf{p}, \mathbf{q}) = \mathbf{p}A\mathbf{q} = \begin{bmatrix} p_1 & p_2 \end{bmatrix} \begin{bmatrix} a_{11} & a_{12} \\ a_{21} & a_{22} \end{bmatrix} \begin{bmatrix} q_1 \\ q_2 \end{bmatrix}$$

$$= a_{11}p_1q_1 + a_{12}p_1q_2 + a_{21}p_2q_1 + a_{22}p_2q_2 . \qquad (5.9)$$

Since

$$p_1 + p_2 = 1 \qquad \text{and} \qquad q_1 + q_2 = 1 \qquad (5.10)$$

we may substitute $p_2 = 1 - p_1$ and $q_2 = 1 - q_1$ into (5.9) to obtain

$$E(\mathbf{p}, \mathbf{q}) = a_{11}p_1q_1 + a_{12}p_1(1 - q_1) + a_{21}(1 - p_1)q_1 + a_{22}(1 - p_1)(1 - q_1) . \qquad (5.11)$$

If we rearrange the terms in Eq. (5.11), we may write

$$E(\mathbf{p}, \mathbf{q}) = [(a_{11} + a_{22} - a_{12} - a_{21})p_1 - (a_{22} - a_{21})]q_1 + (a_{12} - a_{22})p_1 + a_{22}. \qquad (5.12)$$

By examining the coefficient of the q_1 term in (5.12), we see that if we set

$$p_1 = p_1^* = \frac{a_{22} - a_{21}}{a_{11} + a_{22} - a_{12} - a_{21}} , \qquad (5.13)$$

then that coefficient is zero, and Eq. (5.12) reduces to

$$E(\mathbf{p}^*, \mathbf{q}) = \frac{a_{11}a_{22} - a_{12}a_{21}}{a_{11} + a_{22} - a_{12} - a_{21}} . \qquad (5.14)$$

Eq. (5.14) is independent of $\mathbf{q}$. That is, if player R chooses the strategy determined by (5.13), player C cannot change the expected payoff by varying his strategy.

   In a similar manner, it may be verified that if player C chooses the strategy determined by

$$q_1 = q_1^* = \frac{a_{22} - a_{12}}{a_{11} + a_{22} - a_{12} - a_{21}} , \qquad (5.15)$$

then substitution in Eq. (5.12) gives

$$E(\mathbf{p}, \mathbf{q}^*) = \frac{a_{11}a_{22} - a_{12}a_{21}}{a_{11} + a_{22} - a_{12} - a_{21}} \qquad (5.16)$$

Equations (5.14) and (5.16) show that

$$E(\mathbf{p}^*, \mathbf{q}) = E(\mathbf{p}^*, \mathbf{q}^*) = E(\mathbf{p}, \mathbf{q}^*) \qquad (5.17)$$

for all strategies $\mathbf{p}$ and $\mathbf{q}$. Thus, Definition 5.1 is satisfied, so that the strategies determined by Eqs. (5.13) and (5.15) are optimal strategies for players R and C, respectively. Let us summarize this in the following theorem, where we have computed the values of p_2^* and q_2^* from Eq. (5.10).

THEOREM 5.2 *For a 2 × 2 game which is not strictly deter-mined, optimal strategies for players R and C are*

$$\mathbf{p}^* = \left[\frac{a_{22} - a_{21}}{a_{11} + a_{22} - a_{12} - a_{21}} \qquad \frac{a_{11} - a_{12}}{a_{11} + a_{22} - a_{12} - a_{21}} \right]$$

and

$$q^* = \begin{bmatrix} \dfrac{a_{22} - a_{12}}{a_{11} + a_{22} - a_{12} - a_{21}} \\[3mm] \dfrac{a_{11} - a_{21}}{a_{11} + a_{22} - a_{12} - a_{21}} \end{bmatrix}.$$

The value of the game is

$$v = \frac{a_{11}a_{22} - a_{12}a_{21}}{a_{11} + a_{22} - a_{12} - a_{21}}.$$

In order to be complete, we must show that the entries in the vectors p^* and q^* are numbers between 0 and 1. In Exercise 5.7 we ask the reader to show that this is the case as long as the game is not strictly determined.

Equation (5.17) is interesting in that it implies that either player may force the expected payoff to be the value of the game by choosing his optimal strategy, regardless of which strategy the other player chooses. This is not true, in general, for games in which either player has more than two moves.

EXAMPLE 5.3

The Federal government desires to inoculate its citizens against a certain flu virus. The virus has two strains, and it is not known in what proportions the two strains occur in the virus population. Two vaccines have been developed with different effectivenesses against the two strains. Vaccine 1 is 85% effective against strain 1 and 70% effective against strain 2. Vaccine 2 is 60% effective against strain 1 and 90% effective against strain 2. What inoculation policy should the government adopt?

SOLUTION

We may consider this a two-person game in which player R (the government) desires to make the payoff (the fraction of citizens resistant to the virus) as large as possible, and player C (the virus) desires to make the payoff as small as possible. The payoff matrix is

$$
\begin{array}{c}
\text{Strain} \\
\begin{array}{cc} 1 & 2 \end{array}
\end{array}
$$

$$
\text{Vaccine} \begin{array}{c} 1 \\ 2 \end{array} \begin{bmatrix} .85 & .70 \\ .60 & .90 \end{bmatrix}.
$$

It can be seen that this matrix has no saddle points, so that Theorem 5.2 is applicable. Consequently,

$$
p_1^* = \frac{a_{22} - a_{21}}{a_{11} + a_{22} - a_{12} - a_{21}} = \frac{.90 - .60}{.85 + .90 - .70 - .60} = \frac{.30}{.45} = \frac{2}{3}.
$$

$$
p_2^* = 1 - p_1^* = 1 - \frac{2}{3} = \frac{1}{3}.
$$

$$
q_1^* = \frac{a_{22} - a_{12}}{a_{11} + a_{22} - a_{12} - a_{21}} = \frac{.90 - .70}{.85 + .90 - .70 - .60} = \frac{.20}{.45} = \frac{4}{9}.
$$

$$
q_2^* = 1 - q_1^* = 1 - \frac{4}{9} = \frac{5}{9}.
$$

$$
v = \frac{a_{11}a_{22} - a_{12}a_{21}}{a_{11} + a_{22} - a_{12} - a_{21}} = \frac{(.85)(.90) - (.70)(.60)}{.85 + .90 - .70 - .60} = \frac{.345}{.45} = .7666\ldots
$$

Thus, the optimal strategy for the government is to inoculate 2/3 of the citizens with vaccine 1 and 1/3 of the citizens with vaccine 2. This will guarantee that 76.7% of the citizens will be resistant to a virus attack regardless of the distribution of the two strains.

On the other hand, a virus distribution of 4/9 of strain 1 and 5/9 of strain 2 will result in the same 76.7% of resistant citizens regardless of the inoculation strategy adopted by the government.

EXERCISES

5.1 Let a game have payoff matrix

$$
A = \begin{bmatrix} -4 & 6 & -4 & 1 \\ 5 & -7 & 3 & 8 \\ -8 & 0 & 6 & -2 \end{bmatrix}.
$$

(a) If players R and C use strategies

$$\mathbf{p} = \begin{bmatrix} \frac{1}{2} & 0 & \frac{1}{2} \end{bmatrix} \quad \text{and} \quad \mathbf{q} = \begin{bmatrix} \frac{1}{4} \\ \frac{1}{4} \\ \frac{1}{4} \\ \frac{1}{4} \end{bmatrix},$$

respectively, what is the expected payoff of the game?

(b) If player C keeps his strategy fixed as in part (a), what strategy should player R choose to maximize his expected payoff?

(c) If player R keeps his strategy fixed as in part (a), what strategy should Player C choose to minimize the expected payoff to player R?

5.2 Construct a simple example to show that optimal strategies defined in Definition 5.1 are not necessarily unique. For example, find a payoff matrix with several equal saddle points.

5.3 For the strictly determined games with the following payoff matrices, find optimal strategies for the two players and find the values of the games.

(a) $\begin{bmatrix} 5 & 2 \\ 7 & 3 \end{bmatrix}$ (b) $\begin{bmatrix} -3 & -2 \\ 2 & 4 \\ -4 & 1 \end{bmatrix}$ (c) $\begin{bmatrix} 2 & -2 & 0 \\ -6 & 0 & -5 \\ 5 & 2 & 3 \end{bmatrix}$ (d) $\begin{bmatrix} -3 & 2 & -1 \\ -2 & -1 & 5 \\ -4 & 1 & 0 \\ -3 & 4 & 6 \end{bmatrix}$

5.4 For the 2×2 games with the following payoff matrices, find optimal strategies for the two players and find the values of the games.

(a) $\begin{bmatrix} 6 & 3 \\ -1 & 4 \end{bmatrix}$ (b) $\begin{bmatrix} 40 & 20 \\ -10 & 30 \end{bmatrix}$ (c) $\begin{bmatrix} 3 & 7 \\ -5 & 4 \end{bmatrix}$ (d) $\begin{bmatrix} 3 & 5 \\ 5 & 2 \end{bmatrix}$ (e) $\begin{bmatrix} 7 & -3 \\ -5 & -2 \end{bmatrix}$

5.5 Player R has two playing cards: a black ace and a red four. Player C also has two cards: a black two and a red three. Each player secretly selects one of his cards. If both selected cards are the same color, player C pays player R the sum of the face values in dollars. If the cards are different colors, player R pays player C the sum of the face values. What are optimal strategies for both players and what is the value of the game?

5.6 Verify Equations (5.6), (5.7), and (5.8).

5.7 Show that the entries of the optimal strategies p* and q* given in Theorem 5.2 are numbers strictly between zero and one.

The Assignment Problem

A number of facilities are to be assigned
different tasks. Each possible assignment
results in a certain cost. An algorithm
called the Hungarian method is described to
find an assignment with minimum cost.

PREREQUISITES: Matrix notation

INTRODUCTION

A basic problem in operations research is to assign "tasks" to "facilities" on a one-to-one basis in some optimal way. For example, the problem may be to find the best assignment of workers to jobs, sports players to field positions, equipment to construction sites, and so forth. The precise problem requires that there be as many facilities as tasks, say n of each. In this case, there are exactly $n!$ different ways to assign the tasks to the facilities on a one-to-one basis. This follows since there are n ways to assign the first task, $n-1$ ways to assign the second, $n-2$ ways to assign the third, and so on — a total of $n \times (n-1) \times (n-2) \times \cdots \times 3 \times 2 \times 1 = n!$ possible assignments. Among these $n!$ possible assignments, we are to find an optimal one. To define the notion of an optimal assignment precisely we introduce the following quantities:

Let

$$c_{ij} = \text{cost of assigning the } i\text{-th facility the } j\text{-th task}$$

for $i,j = 1, 2, \ldots, n$. The units of c_{ij} might be dollars, miles, hours, etc. — whatever is appropriate to the problem. We define the *cost matrix* to be the $n \times n$ matrix

$$C = \begin{bmatrix} c_{11} & c_{12} & \cdots & c_{1n} \\ c_{21} & c_{22} & \cdots & c_{2n} \\ \vdots & \vdots & & \vdots \\ c_{n1} & c_{n2} & \cdots & c_{nn} \end{bmatrix}.$$

The requirement that each facility be assigned a unique task on a one-to-one basis is equivalent to the condition that no two of the corresponding c_{ij}'s come from the same row or column. This leads to the following definition:

> **DEFINITION 6.1** *Given an $n \times n$ cost matrix C, an* **assignment** *is a set of n of its entries, no two of which come from the same row or column.*

An optimal assignment is then defined as follows:

> **DEFINITION 6.2** *The sum of the n entries of an assignment is called its* **cost**. *An assignment with the smallest possible cost is called an* **optimal assignment**.

The *Assignment Problem* is to find an optimal assignment in a given cost matrix. For example, in assigning n pieces of equipment to n construction sites, c_{ij} could be the distance in miles between the i-th piece of equipment and the j-th construction site. An optimal assignment is one for which the total distance traveled by the n pieces of equipment is a minimum.

EXAMPLE 6.1 A college intends to install air-conditioning in
three of its buildings during a one-week spring break. It invites
three contractors to submit separate bids for the work involved in
each of the three buildings. The bids it receives (in 1000 dollar
units) are listed in the following table:

	BIDS		
	Bldg 1	Bldg 2	Bldg 3
Contractor 1	53	96	37
Contractor 2	47	87	41
Contractor 3	60	92	36

Each contractor can install the air-conditioning for only one building during the one-week period, and so the college must assign a
different contractor to each building. To which building should
each contractor be assigned in order to minimize the sum of the
corresponding bids?

SOLUTION The cost matrix for this problem is the 3×3 matrix

$$\begin{bmatrix} 53 & 96 & 37 \\ 47 & 87 & 41 \\ 60 & 92 & 36 \end{bmatrix}. \tag{6.1}$$

Since there are only six (= 3!) possible assignments, we may solve
this problem by computing the cost of each of them. Below, we have
boxed the entries associated with each of the six assignments and
computed their sums.

$$53 + 87 + 36 = 176$$
(a)

$$53 + 92 + 41 = 186$$
(b)

$$47 + 96 + 36 = 179$$
(c)

$$47 + 92 + 37 = 176$$
(d)

$$60 + 96 + 41 = 197$$
(e)

$$60 + 87 + 37 = 184$$
(f)

Note that the bid totals range from a minimum of $176,000 to a maximum of $197,000. Since the minimum bid total of $176,000 is attained by both assignments (a) and (d), the college should assign the contractors to the buildings in one of the following two ways:

Contractor 1 to Building 1
Contractor 2 to Building 2   or   Contractor 1 to Building 3
Contractor 3 to Building 3   Contractor 2 to Building 1
                                     Contractor 3 to Building 2.

◆—◆—◆

     The method of brute force employed in this example quickly becomes impractical as the size of the cost matrix increases. For example, for a 10×10 cost matrix there are a total of 3,628,800 ($= 10!$) assignments possible. In the next section we describe a practical method for solving any assignment problem.

THE HUNGARIAN METHOD

Suppose a particular assignment problem has the cost matrix

$$
\begin{bmatrix}
0 & 5 & 7 & 0 & 3 \\
0 & 4 & 0 & 2 & 5 \\
6 & 0 & 3 & 7 & 7 \\
7 & 9 & 4 & 0 & 0 \\
3 & 2 & 0 & 0 & 1
\end{bmatrix}. \tag{6.2}
$$

Notice that all of the entries of this cost matrix are nonnegative and that it contains many zero entries. Notice, also, that we can find an assignment consisting entirely of zero entries, namely

$$
\begin{bmatrix}
\boxed{0} & 5 & 7 & 0 & 3 \\
0 & 4 & \boxed{0} & 2 & 5 \\
6 & \boxed{0} & 3 & 7 & 7 \\
7 & 9 & 4 & 0 & \boxed{0} \\
3 & 2 & 0 & \boxed{0} & 1
\end{bmatrix}. \tag{6.3}
$$

This assignment must be optimal since its cost is zero and since it is impossible to find an assignment with a cost less than zero if all of the entries of the cost matrix are nonnegative.
     Very few assignment problems are as easy to solve as the one above. However, the following theorem leads to a method of con-

verting an arbitrary assignment problem to one which can be solved as easily.

> **THEOREM 6.1** *If a number is added to or subtracted from all of the entries of any one row or column of a cost matrix, then an optimal assignment for the resulting cost matrix is also an optimal assignment for the original cost matrix.*

To see why this theorem is true, suppose that five is added to each entry of the second row of a given cost matrix. Since each assignment contains exactly one entry from the second row, it follows that the cost of each assignment for the new matrix is exactly five more than the cost of the corresponding assignment for the original matrix. Thus, corresponding assignments preserve their ordering with respect to cost, and so an optimal assignment of either matrix corresponds to an optimal assignment of the other. A similar argument holds if a number is added to any column of the cost matrix, or if subtraction rather than addition is used.

We now introduce the *Hungarian method,* which is a procedure for applying this theorem to a given cost matrix and obtaining one with nonnegative entries which contains an assignment consisting entirely of zero entries. Such an assignment (called an *optimal assignment of zeroes*) will then be an optimal assignment for the original problem. The Hungarian method is outlined in Fig. 6.1. The first two steps use Theorem 6.1 to generate a cost matrix with nonnegative entries and with at least one zero entry in each row and column. The last three steps are applied iteratively as many times as necessary to generate a cost matrix which contains an optimal assignment of zeroes. In Exercise 6.8, we ask the reader to show that each time Step 5 is applied, the sum of the entries of the new cost matrix generated is strictly less than the sum of the entries of the previous cost matrix. This guarantees that the iterative process cannot continue indefinitely.

THE HUNGARIAN METHOD

STEPS	REMARKS
1. Subtract the smallest entry in each row from all of the entries of its row.	After this step, each row has at least one zero entry and all other entries are nonnegative.
2. Subtract the smallest entry in each column from all of the entries of its column.	After this step, each row *and* column has at least one zero entry and all other entries are nonnegative.
3. Draw lines through appropriate rows and columns so that all the zero entries of the cost matrix are covered and the *minimum* number of such lines is used.	There may be several ways to do this. The important thing is that the *minimum* number of lines be used. Algorithms suitable for computer coding are available for this; however, for small values of n trial and error will suffice.
4. **Test for Optimality.** (i) If the minimum number of covering lines is n, an optimal assignment of zeroes is possible and we are finished. (ii) If the minimum number of covering lines is less than n, an optimal assignment of zeroes is not yet possible. Proceed to Step 5.	See Exercise 6.7 for a justification of this test. If the test is affirmative, a judicious search will produce a set of n zero entries no two of which lie in the same row or column. Algorithms, which we shall not discuss, are available to systematically find such an optimal assignment of zeroes.
5. Determine the smallest entry not covered by any line. Subtract this entry from all uncovered entries and then add it to all entries covered by both a horizontal and a vertical line. Return to Step 3.	This step is equivalent to applying Theorem 6.1 by subtracting the smallest uncovered entry from each uncovered row and then adding it to each covered column.

Figure 6.1

EXAMPLE 6.2 A construction company has four large bulldozers located at four different garages. The four bulldozers are to be moved to four different construction sites. The distances in miles between the various bulldozers and construction sites are given in the following table:

		Construction Site			
		1	2	3	4
Bulldozer	1	90	75	75	80
	2	35	85	55	65
	3	125	95	90	105
	4	45	110	95	115

How should the bulldozers be moved to the construction sites in order to minimize the total distance traveled?

SOLUTION Matrix (6.4) is the cost matrix for this problem. The steps of the Hungarian method are as follows:

Step 1. Subtract 75 from the first row of matrix (6.4), subtract 35 from its second row, subtract 90 from its third row, and subtract 45 from its fourth row to obtain matrix (6.5).

$$\begin{bmatrix} 90 & 75 & 75 & 80 \\ 35 & 85 & 55 & 65 \\ 125 & 95 & 90 & 105 \\ 45 & 110 & 95 & 115 \end{bmatrix} \quad (6.4)$$

Step 2. The first three columns of matrix (6.5) already contain zero entries, therefore we need only subtract 5 from its fourth column. The result is matrix (6.6).

$$\begin{bmatrix} 15 & 0 & 0 & 5 \\ 0 & 50 & 20 & 30 \\ 35 & 5 & 0 & 15 \\ 0 & 65 & 50 & 70 \end{bmatrix} \quad (6.5)$$

Step 3. Cover the zero entries of matrix (6.6) with a minimum number of vertical and horizontal lines. This may be done by first trying to cover the zeroes with one line, then with two, and finally with three. The indicated covering is not unique. See Exercise 6.3 for an alternate covering.

$$\begin{bmatrix} -15 & 0 & 0 & 0 \\ 0 & 50 & 20 & 25 \\ -35 & 5 & 0 & 10 \\ 0 & 65 & 50 & 65 \end{bmatrix} \quad (6.6)$$

Step 4. Since the minimum number of lines used in Step 3 is three, an optimal assignment of zeroes is not yet possible.

Step 5. Subtract 20, the smallest uncovered entry of matrix (6.6), from each of its uncovered entries and add it to the two entries covered twice with lines. The result is matrix (6.7)

$$\begin{bmatrix} -35 & 0 & 0 & 0 \\ 0 & 30 & 0 & 5 \\ 55 & 5 & 0 & 10 \\ 0 & 45 & 30 & 45 \end{bmatrix} \quad (6.7)$$

Step 3. Cover the zero entries of matrix (6.7) with a minimum number of vertical and horizontal lines.

Step 4. Since the minimum number of lines is still three, an optimal assignment of zeroes is not yet possible.

Step 5. Subtract 5, the smallest uncovered entry of matrix (6.7), from each of its uncovered entries and add it to the two entries covered twice with lines. The result is matrix (6.8).

$$\begin{bmatrix} -40 & 0 & 5 & 0 \\ 0 & 25 & 0 & 0 \\ 55 & 0 & 0 & 5 \\ 0 & 40 & 30 & 40 \end{bmatrix} \quad (6.8)$$

Step 3. Cover the zero entries of matrix (6.8) with a minimum number of vertical and horizontal lines.

Step 4. Since the zero entries of matrix (6.8) cannot be covered with fewer than four lines, it must contain an optimal assignment of zeroes.

By trial and error, we can find the following two optimal assignments of zeroes in matrix (6.8):

$$\begin{bmatrix} 40 & 0 & 5 & \boxed{0} \\ 0 & 25 & \boxed{0} & 0 \\ 55 & \boxed{0} & 0 & 5 \\ \boxed{0} & 40 & 30 & 40 \end{bmatrix} \qquad \begin{bmatrix} 40 & \boxed{0} & 5 & 0 \\ 0 & 25 & 0 & \boxed{0} \\ 55 & 0 & \boxed{0} & 5 \\ \boxed{0} & 40 & 30 & 40 \end{bmatrix}. \quad (6.9)$$

(a)   (b)

Assignment (a) leads to the following movement of bulldozers to construction sites:

Bulldozer 1 to construction site 4
      "   2 "   "   " 3
      "   3 "   "   " 2
      "   4 "   "   " 1

From the table in the statement of the problem, the corresponding minimum distance traveled is

$$80 + 55 + 95 + 45 = 275 \text{ miles.}$$

Similarly, assignment (b) leads to the alternate solution

Bulldozer 1 to construction site 2
      "   2 "   "   " 4
      "   3 "   "   " 3
      "   4 "   "   " 1

with the same minimum distance traveled:

$$75 + 65 + 90 + 45 = 275 \text{ miles.}$$

◆ ◆ ◆

Before presenting another example, we mention that the assignment problems and associated cost matrices that can be solved by the Hungarian method must satisfy the following three conditions:

1. *The cost matrix must be square.* See Example 6.3 and Exercise 6.5 for assignment problems in which the cost matrix is not initially square.

2. *The entries of the cost matrix should be integers.* For hand calculations this is more a convenience than anything else. But for machine calculations, this allows one to use exact integer arithmetic and avoid roundoff error. For practical problems, noninteger entries can always be converted to integer entries by multiplying the cost matrix by a suitable power of ten.

3 . *The problem must be one of minimization.* The problem of maximizing the sum of entries of a cost matrix is easily converted to one of minimizing the sum by multiplying each entry of the cost matrix by -1.

EXAMPLE 6.3 (THE BRIDE-GROOM PROBLEM)

A marriage broker has four female clients and five male clients who desire to be married. She ranks the possible matchings between her clients on a scale of zero to ten: zero for the poorest match and ten for the best match. Her rankings are tabulated as follows:

		Prospective Grooms				
		Bob	Tom	Joe	Hal	Don
Prospective Brides	Sue	7	4	7	3	10
	Ann	5	9	3	8	7
	Bea	3	5	6	2	9
	Fay	6	5	0	4	8

How should she match her clients in order the maximize the sum of the rankings of the matches?

SOLUTION
Since there is one more prospective groom than prospective bride, one of the prospective grooms cannot be matched. To handle this mathematically, we introduce a "dummy" prospective bride whose match with any of the five prospective grooms has a rank of zero. The prospective groom matched with the dummy bride is then, in reality, the groom not matched. We therefore add a row of zeroes to the table above corresponding to the dummy bride and in this way are led to the following square 5 × 5 cost matrix:

$$\begin{bmatrix} 7 & 4 & 7 & 3 & 10 \\ 5 & 9 & 3 & 8 & 7 \\ 3 & 5 & 6 & 2 & 9 \\ 6 & 5 & 0 & 4 & 8 \\ 0 & 0 & 0 & 0 & 0 \end{bmatrix}. \qquad (6.10)$$

The problem as stated is one of maximizing a sum. We convert it to one of minimizing a sum by multiplying each entry of matrix (6.10) by -1 to obtain

$$\begin{bmatrix} -7 & -4 & -7 & -3 & -10 \\ -5 & -9 & -3 & -8 & -7 \\ -3 & -5 & -6 & -2 & -9 \\ -6 & -5 & 0 & -4 & -8 \\ 0 & 0 & 0 & 0 & 0 \end{bmatrix}. \qquad (6.11)$$

We now apply the Hungarian method to matrix (6.11) to find an optimal assignment.

Step 1. Subtract -10 from the first row of matrix (6.11), subtract -9 from its second row, subtract -9 from its third row, and subtract -8 from its fourth row to obtain matrix (6.12).

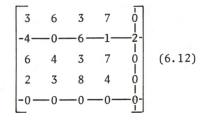

(6.12)

Step 2. Since all of the columns of matrix (6.12) contain zero entries, this step is not needed.

Step 3. Cover the zero entries of matrix (6.12) with a minimum number of vertical and horizontal lines.

Step 4. Since the minimum number of lines used in Step 3 is three, an optimal assignment of zeroes is not possible.

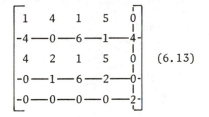

(6.13)

Step 5. Subtract 2, the smallest uncovered entry of matrix (6.12), from each of its uncovered entries and add it to the two entries covered twice with lines. The result is matrix (6.13).

Step 3. Cover the zero entries of matrix (6.13) with a minimum number of vertical and horizontal lines.

Step 4. Since the minimum number of lines is four, an optimal assignment of zeroes is not yet possible.

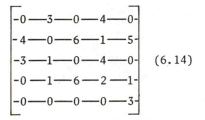

(6.14)

Step 5. Subtract 1, the smallest uncovered entry of matrix (6.13), from each of its uncovered entries and add it to the three entries covered twice with lines. The result is matrix (6.14).

Step 3. Cover the zero entries of matrix (6.14) with a minimum number of vertical and horizontal lines.

Step 4. Since the zero entries of
matrix (6.14) cannot be covered
with fewer than five lines, it
must contain an optimal assignment
of zeroes. Such an assignment is
given in matrix (6.15).

$$\begin{bmatrix} 0 & 3 & \boxed{0} & 4 & 0 \\ 4 & \boxed{0} & 6 & 1 & 5 \\ 3 & 1 & 0 & 4 & \boxed{0} \\ \boxed{0} & 1 & 6 & 2 & 1 \\ 0 & 0 & 0 & \boxed{0} & 3 \end{bmatrix} \quad (6.15)$$

The optimal assignment indicated in (6.15) leads to the following
bride-groom matchings:

Sue - Joe	(rank = 7)
Ann - Tom	(rank = 9)
Bea - Don	(rank = 9)
Fay - Bob	(rank = 6)
Hal unmatched,	

and the resulting maximum ranking sum is $7 + 9 + 9 + 6 = 31$. In Exercise 6.4 we ask the reader to find an alternate optimal assignment of zeroes in matrix (6.15) and to verify that it leads to the same maximum ranking sum.

EXERCISES

6.1 Find an optimal assignment and corresponding cost for each of the following three cost matrices using the Hungarian method.

(a) $\begin{bmatrix} 17 & 4 & 10 \\ 15 & 5 & 8 \\ 18 & 7 & 11 \end{bmatrix}$
(b) $\begin{bmatrix} 3 & -2 & 0 & 1 \\ 5 & 3 & -3 & 4 \\ 2 & 7 & 5 & 3 \\ 5 & -2 & 0 & 1 \end{bmatrix}$
(c) $\begin{bmatrix} 12 & 9 & 7 & 7 & 10 \\ 15 & 11 & 8 & 13 & 14 \\ 9 & 6 & 5 & 12 & 12 \\ 6 & 9 & 13 & 7 & 10 \\ 8 & 13 & 12 & 9 & 13 \end{bmatrix}$

6.2 Solve Example 6.1 by the Hungarian method.

6.3 Find three lines which cover the zero entries of matrix (6.6) in Example 6.2 different from the set given. Then use the Hungarian method and verify that it leads to the same optimal assignments of zeroes given in (6.9).

6.4 Find an optimal assignment of zeroes in matrix (6.15) of Example 6.3 different from the one given. Verify that the new assignment leads to the same cost as the one in the example.

6.5 A coin dealer is to sell four coins through a mail auction. Bids are received for each of the four coins from five bidders with instructions from each bidder that at most one of his bids is to be honored. The bids are given in the following table:

	BIDS			
	Coin 1	Coin 2	Coin 3	Coin 4
Bidder 1	$150	$ 65	$210	$135
Bidder 2	175	75	230	155
Bidder 3	135	85	200	140
Bidder 4	140	70	190	130
Bidder 5	170	50	200	160

(a) How should the dealer assign the four coins in order to maximize the sum of the resulting bids? (Notice that a "dummy" coin must be added to produce a square cost matrix. Whichever bidder receives the dummy coin does not receive any real coin.)

(b) Suppose Bidder 2 instructed the dealer that at most two of his bids are to be honored. How can the problem be modified and what is an optimal assignment for the new problem?

6.6 A Little League manager has nine players to assign to the nine positions of a baseball team. The manager ranks each of the players on a scale from 0 to 25 for each of the positions, taking into account both the ability of the player in the position and the importance of the position to the game. Referring to the table below, how should the manager assign the nine players in order to maximize the sum of the rankings?

		PLAYER								
		Sam	Jill	John	Liz	Ann	Lois	Pete	Alex	Herb
Position	P	20	15	10	10	17	23	25	5	15
	C	10	10	12	15	9	7	8	7	8
	1B	12	9	9	10	10	5	7	13	9
	2B	13	14	10	15	15	5	8	20	10
	3B	12	13	10	15	14	5	9	20	10
	SS	15	14	15	16	15	5	10	20	10
	LF	7	9	12	12	7	6	7	15	12
	CF	5	6	8	8	5	4	5	10	7
	RF	5	6	8	8	5	4	5	10	7

6.7 Justify the Test for Optimality in Step 4 of the Hungarian method as follows:

(a) Show that the n zero entries of an optimal assignment of zeroes cannot be covered with fewer than n lines. (Thus, if fewer than n lines can cover all of the zero entries of a matrix, then it cannot contain an optimal assignment of zeroes.)

(b) Show that the zero entries of a matrix that does not contain an optimal assignment of zeroes can always be covered with fewer than n lines. (Thus, if the zero entries of a matrix cannot be covered with fewer than n lines, then it must contain an optimal assignment of zeroes.)

Remark: Part (b) is quite difficult and should be regarded as a research project. It was first proved in 1931 by two Hungarian mathematicians: D. König and E. Egervary. It is after them that the Hungarian Method is named.

6.8 (a) Let the zero entries of an $n \times n$ nonnegative cost matrix C be covered with m lines, with $m < n$. Let $a > 0$ be the smallest uncovered entry and let C' be the cost matrix obtained by applying Step 5 of the Hungarian method to C. Show that

(Sum of all entries of C) - (Sum of all entries of C')

$= an(n - m)$.

Hint: Notice that Step 5 is equivalent to adding a to every entry of C covered by one line, adding $2a$ to every entry covered by two lines, and subtracting a from every entry.

(b) Use the result of part (a) to show that the Hungarian method will produce a cost matrix with an optimal assignment of zeroes in a finite number of steps.

7

Markov Chains

A general model of a system which moves from state to state is described and applied to several concrete problems. It is shown that such systems tend to a steady-state configuration for large times.

PREREQUISITES: Linear systems
Matrices
Intuitive understanding of limits

INTRODUCTION

Suppose a physical or mathematical system is such that at any moment it can occupy one of a finite number of states. For example, the weather in a certain city could be in one of three possible states: sunny, cloudy, or raining. Or an individual could be in one of four possible emotional states: happy, sad, angry, or apprehensive. Suppose such a system moves with time from one state to another. Let us construct a schedule of observation times and keep a record of the states of the system at these times. If we find that the transition from one state to another is not predetermined, but rather can only be specified in terms of certain probabilities which depend on the previous history of the system, then the process is called a *stochastic process*. If, in addition, these transition probabilities depend only on the immediate history of the system, that is, if the state of the system at any observation is dependent only on its state at the immediately preceeding observation, then the process is called a *Markov process* or a *Markov chain*.

Let the system under observation have k possible states, which we label as states $1, 2, \ldots, k$. We define the following quantities to describe the transitions of the system from one state to another:

> **DEFINITION 7.1** *The* **transition probability** p_{ij} *(i, j = 1, 2, ..., k) is the probability that if the system is in state j at any one observation, it will be in state i at the next observation.*

For example, if state 2 corresponds to a rainy day in Detroit, and state 3 corresponds to a cloudy day, then p_{32} is the probability that the weather in Detroit changes from rainy to cloudy in two successive days.

As probabilities, the numbers p_{ij} must all lie in the interval $[0, 1]$. Furthermore, for any fixed $j = 1, 2, \ldots, k$, we must have

$$p_{1j} + p_{2j} + \cdots + p_{kj} = 1, \tag{7.1}$$

which expresses the fact that if the system is in state j at one observation, it will with certainty be in one of the k states at the next observation.

With these k^2 transition probabilities, we may form a $k \times k$ matrix $P = [p_{ij}]$, called the *transition matrix of the Markov process*. Equation (7.1) expresses the fact that the sum of the entries in each column of P is one. More generally, we have the following definition:

> **DEFINITION 7.2** *A* **transition matrix** $P = [p_{ij}]$ *is any square matrix with nonnegative entries, all of whose column sums are one.*

Thus, any Markov process determines a transition matrix. Transition matrices are also called *Markov matrices*, *probability matrices*, or *stochastic matrices*.

EXAMPLE 7.1 By reviewing its donation records, the alumni office of a college finds that 80% of its alumni who contribute to the annual fund one year will also contribute the next year, and 30% of those who do not contribute one year will contribute the next. This may be described in terms of a Markov process with two states: state 1 corresponds to an alumnus giving a donation in any one year, and state 2 corresponds to the alumnus not giving a donation in that year. The transition matrix of the process is

$$P = \begin{bmatrix} .8 & .3 \\ .2 & .7 \end{bmatrix}.$$

Our objective in this chapter is to predict the state of a system described by a Markov process at future observation times. However, since the system is not deterministic, our predictions will be in terms of probabilities. To this end we introduce the following definition:

> **DEFINITION 7.3** *A* **probability vector** *is a column vector with nonnegative entries whose sum is one.*

We may then specify a future observation of a system described by a Markov process in terms of a probability vector as follows:

> **DEFINITION 7.4** *The probability vectors* $x^{(n)}$ *for* $n = 0, 1,$ *... are said to be the* **state vectors** *of a Markov process if the i-th component* $x_i^{(n)}$ *of* $x^{(n)}$ *is the probability that the system is in the i-th state at the n-th observation.*

In particular, the state vector $x^{(0)}$ is called the *initial state vector* of the Markov process. Using the next theorem, we shall see that the initial state vector determines all future state vectors.

> **THEOREM 7.1** *If P is the transition matrix of a Markov process and* $x^{(n)}$ *is the state vector at the n-th observation, then* $x^{(n+1)} = P x^{(n)}$.

The proof of this theorem involves ideas from probability theory and will not be given here. From this theorem it follows that

$$x^{(1)} = Px^{(0)}$$
$$x^{(2)} = Px^{(1)} = P^2 x^{(0)}$$
$$x^{(3)} = Px^{(2)} = P^3 x^{(0)}$$
$$\vdots$$
$$x^{(n)} = Px^{(n-1)} = P^n x^{(0)}.$$

In this way, the initial state vector $x^{(0)}$ and the transition matrix P determine $x^{(n)}$ for $n = 1, 2, \ldots$.

EXAMPLE 7.1 (REVISITED)   The transition matrix in Example 7.1 was

$$P = \begin{bmatrix} .8 & .3 \\ .2 & .7 \end{bmatrix}.$$

Let us construct the probable future donation record of a new graduate who did not give a donation in the initial year after his graduation. In this way, the system is initially in state 2 with certainty, and so the initial state vector is

$$x^{(0)} = \begin{bmatrix} 0 \\ 1 \end{bmatrix}.$$

From Theorem 7.1, we then have

$$x^{(1)} = Px^{(0)} = \begin{bmatrix} .8 & .3 \\ .2 & .7 \end{bmatrix} \begin{bmatrix} 0 \\ 1 \end{bmatrix} = \begin{bmatrix} .3 \\ .7 \end{bmatrix}$$

$$x^{(2)} = Px^{(1)} = \begin{bmatrix} .8 & .3 \\ .2 & .7 \end{bmatrix} \begin{bmatrix} .3 \\ .7 \end{bmatrix} = \begin{bmatrix} .45 \\ .55 \end{bmatrix}$$

$$x^{(3)} = Px^{(2)} = \begin{bmatrix} .8 & .3 \\ .2 & .7 \end{bmatrix} \begin{bmatrix} .45 \\ .55 \end{bmatrix} = \begin{bmatrix} .525 \\ .475 \end{bmatrix}.$$

Thus, after three years, the alumnus can be expected to make a donation with probability .525. Beyond three years, we find the following state vectors (to three decimal places):

$$x^{(4)} = \begin{bmatrix} .563 \\ .438 \end{bmatrix}, \quad x^{(5)} = \begin{bmatrix} .581 \\ .419 \end{bmatrix}, \quad x^{(6)} = \begin{bmatrix} .591 \\ .409 \end{bmatrix}, \quad x^{(7)} = \begin{bmatrix} .595 \\ .405 \end{bmatrix}$$

$$x^{(8)} = \begin{bmatrix} .598 \\ .402 \end{bmatrix}, \quad x^{(9)} = \begin{bmatrix} .599 \\ .401 \end{bmatrix}, \quad x^{(10)} = \begin{bmatrix} .599 \\ .401 \end{bmatrix}, \quad x^{(11)} = \begin{bmatrix} .600 \\ .400 \end{bmatrix}.$$

For all n beyond eleven, we have $x^{(n)} = \begin{bmatrix} .600 \\ .400 \end{bmatrix}$ to three decimal places. In other words, the state vectors converge to a fixed vector as the number of observations increases. We shall discuss this further in the next section.

EXAMPLE 7.2

A car rental agency has three rental locations, which we label as locations 1, 2, and 3. A customer may rent a car from any of the three locations and return the car to any of the three locations. The manager finds that the customers return the cars to the various locations according to the following probabilities:

Rented from location

$$\begin{array}{ccc} 1 & 2 & 3 \end{array}$$

$$\begin{bmatrix} .8 & .3 & .2 \\ .1 & .2 & .6 \\ .1 & .5 & .2 \end{bmatrix} \begin{array}{l} 1 \\ 2 \\ 3 \end{array} \quad \begin{array}{l} Returned \\ to \\ location \end{array}$$

This matrix is the transition matrix of the system considered as a Markov process. The i-th component of a state vector is the probability that a car is returned to the i-th location. In the table below, we list the state vectors $x^{(n)}$ for $n = 0, 1, \ldots, 11$ when a car is initially rented from location 2.

$x^{(n)}$ \ n	0	1	2	3	4	5	6	7	8	9	10	11
$x_1^{(n)}$	0	.300	.400	.477	.511	.533	.544	.550	.553	.555	.556	.557
$x_2^{(n)}$	1	.200	.370	.252	.261	.240	.238	.233	.232	.231	.230	.230
$x_3^{(n)}$	0	.500	.230	.271	.228	.227	.219	.217	.215	.214	.214	.213

For all values of n greater than eleven, all state vectors are equal to $x^{(11)}$ to three decimal places.

Two things should be observed in this example. First, it was not necessary to know how long a customer kept his car. That is, in a Markov process the time period between observations need not be regular. Second, the state vectors approach a fixed vector as n increases, just as in the first example.

EXAMPLE 7.3

A traffic officer is assigned to control the traffic at the eight intersections indicated in Fig. 7.1. He is instructed to remain at each intersection for an hour and then to either remain at that same intersection or move to a neighboring intersection. To avoid establishing a pattern, he is told to choose his new intersection on a random basis, with each possible choice equally likely. For example, if he is at intersection 5, his next intersection can be 2, 4, 5, or 8, each with probability 1/4. Every day he starts at the location where he stopped the day before. As a Markov process, we may construct the following transition matrix:

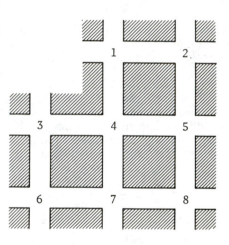

Figure 7.1

Old Intersection

	1	2	3	4	5	6	7	8	
	1/3	1/3	0	1/5	0	0	0	0	1
	1/3	1/3	0	0	1/4	0	0	0	2
	0	0	1/3	1/5	0	1/3	0	0	3
	1/3	0	1/3	1/5	1/4	0	1/4	0	4
	0	1/3	0	1/5	1/4	0	0	1/3	5
	0	0	1/3	0	0	1/3	1/4	0	6
	0	0	0	1/5	0	1/3	1/4	1/3	7
	0	0	0	0	1/4	0	1/4	1/3	8

New Intersection

If the officer initially begins at intersection 5, his probable locations, hour by hour, are given by the following state vectors written in tabular form.

$x^{(n)}$ \ n	0	1	2	3	4	5	10	15	20	22
$x_1^{(n)}$	0	.000	.133	.116	.130	.123	.113	.109	.108	.107
$x_2^{(n)}$	0	.250	.146	.163	.140	.138	.115	.109	.108	.107
$x_3^{(n)}$	0	.000	.050	.039	.067	.073	.100	.106	.107	.107
$x_4^{(n)}$	0	.250	.113	.187	.162	.178	.178	.179	.179	.179
$x_5^{(n)}$	1	.250	.279	.190	.190	.168	.149	.144	.143	.143
$x_6^{(n)}$	0	.000	.000	.050	.056	.074	.099	.105	.107	.107
$x_7^{(n)}$	0	.000	.133	.104	.131	.125	.138	.141	.143	.143
$x_8^{(n)}$	0	.250	.146	.152	.124	.121	.108	.107	.107	.107

For all values of n greater than 22, all state vectors are equal to $x^{(22)}$ to three decimal places. So as with the first two examples, the state vectors approach a fixed vector as n increases.

LIMITING BEHAVIOR OF THE STATE VECTORS

In our three examples, we saw that the state vectors approached some fixed vector as the number of observations increased. We now ask whether the state vectors always approach a fixed vector in a Markov process. A simple example shows that this is not the case:

EXAMPLE 7.4

Let $P = \begin{bmatrix} 0 & 1 \\ 1 & 0 \end{bmatrix}$ and $x^{(0)} = \begin{bmatrix} 1 \\ 0 \end{bmatrix}$. Then, since $P^2 = I$ and $P^3 = P$, we have that

$$x^{(0)} = x^{(2)} = x^{(4)} = \cdots = \begin{bmatrix} 1 \\ 0 \end{bmatrix}$$

and

$$x^{(1)} = x^{(3)} = x^{(6)} = \cdots = \begin{bmatrix} 0 \\ 1 \end{bmatrix}.$$

This system oscillates indefinitely between the two state vectors $\begin{bmatrix} 1 \\ 0 \end{bmatrix}$ and $\begin{bmatrix} 0 \\ 1 \end{bmatrix}$, and so does not approach any fixed vector.

◆━◆━◆

However, if we impose a mild condition on the transition matrix, we can show that a fixed limiting state vector is approached. This condition is described by the following definition:

DEFINITION 7.5 *A transition matrix is* **regular** *if some integer power of it has all positive entries.*

Thus, for a regular transition matrix P, there is some positive integer m such that all entries of P^m are positive. This is the case with the transition matrices of Examples 7.1 and 7.2 for $m = 1$. In Example 7.3, it turns out that P^4 has all positive entries. Consequently, in all three examples the transition matrices are regular.

A Markov process which is governed by a regular transition matrix is called a *regular Markov process*. We shall see that every regular Markov process has a fixed state vector q such that $P^n x^{(0)}$ approaches q as n increases for any choice of $x^{(0)}$. This result is of major importance in the theory of Markov chains. It is based on the following theorem:

THEOREM 7.2 *If P is a regular transition matrix, then as $n \to \infty$*

$$P^n \longrightarrow \begin{bmatrix} q_1 & q_1 & \cdots & q_1 \\ q_2 & q_2 & \cdots & q_2 \\ \vdots & \vdots & & \vdots \\ q_k & q_k & \cdots & q_k \end{bmatrix}$$

where the q_i are positive numbers such that $q_1 + q_2 + \cdots + q_k = 1$.

We shall not prove this theorem here. The interested reader is referred to a more specialized text, such as "Finite Markov Chains" by J. Kemeny and J. Snell (Springer-Verlag New York Inc., 1976).

Let us set

$$Q = \begin{bmatrix} q_1 & q_1 & \cdots & q_1 \\ q_2 & q_2 & \cdots & q_2 \\ \vdots & \vdots & & \vdots \\ q_k & q_k & \cdots & q_k \end{bmatrix} \quad \text{and} \quad q = \begin{bmatrix} q_1 \\ q_2 \\ \vdots \\ q_k \end{bmatrix}.$$

Thus Q is a transition matrix, all of whose columns are equal to the probability vector q. Q has the property that if x is any probability vector, then

$$Qx = \begin{bmatrix} q_1 & q_1 & \cdots & q_1 \\ q_2 & q_2 & \cdots & q_2 \\ \vdots & \vdots & & \vdots \\ q_k & q_k & \cdots & q_k \end{bmatrix} \begin{bmatrix} x_1 \\ x_2 \\ \vdots \\ x_k \end{bmatrix} = \begin{bmatrix} q_1 x_1 + q_1 x_2 + \cdots + q_1 x_k \\ q_2 x_1 + q_2 x_2 + \cdots + q_2 x_k \\ \vdots \\ q_k x_1 + q_k x_2 + \cdots + q_k x_k \end{bmatrix}$$

$$= (x_1 + x_2 + \cdots + x_k) \begin{bmatrix} q_1 \\ q_2 \\ \vdots \\ q_k \end{bmatrix} = (1)q = q.$$

That is, Q transforms any probability vector x into the fixed probability vector q. From this, we can conclude the following:

THEOREM 7.3 *If P is a regular transition matrix and* x *is any probability vector, then as* $n \to \infty$

$$P^n x \to \begin{bmatrix} q_1 \\ q_2 \\ \vdots \\ q_k \end{bmatrix} = q,$$

> *where* q *is a fixed probability vector independent of* n
> *all of whose entries are positive.*

This follows since from Theorem 7.2 we have that $P^n \to Q$ as $n \to \infty$. This in turn implies that $P^n x \to Qx = q$ as $n \to \infty$. $\square$

Thus, for a regular Markov process, the system eventually approaches a fixed state vector q. The vector q is called the *steady-state vector* of the regular Markov chain.

For systems with many states, usually the most efficient technique of computing the steady-state vector q is simply to calculate $P^n x$ for some large n. Our three examples illustrate this procedure. Each is a regular Markov process, so that convergence to a steady-state vector is assured. Another way of computing the steady-state vector is to make use of the following theorem:

> **THEOREM 7.4** *The steady-state vector* q *of a regular*
> *transition matrix* P *is the unique probability vector*
> *which satisfies the equation* $Pq = q$.

To see this, consider the matrix identity $P P^n = P^{n+1}$. By Theorem 7.1, both P^n and P^{n+1} approach Q as $n \to \infty$. Thus we have $PQ = Q$. Any one column of this matrix equation gives $Pq = q$. To show that q is the only probability vector which satisfies this equation, suppose r is another probability vector such that $Pr = r$. Then also $P^n r = r$ for $n = 1, 2, \ldots$. Letting $n \to \infty$, Theorem 7.3 leads to $q = r$. $\square$

Theorem 7.4 may also be expressed by the statement that the homogeneous linear system

$$(I - P)q = 0$$

has a unique vector solution q such that $q_1 + q_2 + \cdots + q_k = 1$. Let us apply this technique to the computation of the steady-state vectors for our three examples.

EXAMPLE 7.1 (REVISITED)

In Example 7.1, the transition matrix was

$$P = \begin{bmatrix} .8 & .3 \\ .2 & .7 \end{bmatrix},$$

so that the linear system $(I - P)q = 0$ is

$$\begin{bmatrix} .2 & -.3 \\ -.2 & .3 \end{bmatrix} \begin{bmatrix} q_1 \\ q_2 \end{bmatrix} = \begin{bmatrix} 0 \\ 0 \end{bmatrix}. \tag{7.2}$$

This leads to the single independent equation

$$.2q_1 - .3q_2 = 0$$

or

$$q_1 = 1.5q_2.$$

Thus, setting $q_2 = s$, any solution of (7.2) is of the form $q = s \begin{bmatrix} 1.5 \\ 1 \end{bmatrix}$

where s is an arbitrary constant. To make q a probability vector we set $s = 1/(1.5 + 1) = .4$. Consequently,

$$q = \begin{bmatrix} .6 \\ .4 \end{bmatrix}$$

is the steady-state vector of this regular Markov process. This means that over the long run, 60% of the alumni will give a donation in any one year, and 40% will not. Observe that this agrees with the result obtained numerically in Example 7.1.

EXAMPLE 7.2 (REVISITED)
In Example 7.2, the transition matrix was

$$P = \begin{bmatrix} .8 & .3 & .2 \\ .1 & .2 & .6 \\ .1 & .5 & .2 \end{bmatrix},$$

so that the linear system $(I - P)q = 0$ is

$$\begin{bmatrix} .2 & -.3 & -.2 \\ -.1 & .8 & -.6 \\ -.1 & -.5 & .8 \end{bmatrix} \begin{bmatrix} q_1 \\ q_2 \\ q_3 \end{bmatrix} = \begin{bmatrix} 0 \\ 0 \\ 0 \end{bmatrix}.$$

The reduced row-echelon form of the coefficient matrix is (verify)

$$\begin{bmatrix} 1 & 0 & -34/13 \\ 0 & 1 & -14/13 \\ 0 & 0 & 0 \end{bmatrix},$$

so that the original linear system is equivalent to the system

$$q_1 = (34/13)q_3$$
$$q_2 = (14/13)q_3 \; .$$

Setting $q_3 = s$, any solution of the linear system is of the form

$$\mathbf{q} = s \begin{bmatrix} 34/13 \\ 14/13 \\ 1 \end{bmatrix}.$$

To make this a probability vector, we set $s = \dfrac{1}{\dfrac{34}{13} + \dfrac{14}{13} + 1} = \dfrac{13}{61}$. Thus the steady-state vector of the system is

$$\mathbf{q} = \begin{bmatrix} 34/61 \\ 14/61 \\ 13/61 \end{bmatrix} = \begin{bmatrix} .5573... \\ .2295... \\ .2131... \end{bmatrix}.$$

This agrees with the result obtained numerically in Example 7.2. The entries of $\mathbf{q}$ give the long-run probabilities that any one car will be returned to location 1, 2, or 3 respectively. If the car rental agency has a fleet of 1000 cars, it should thus design its facilities so that there are at least 558 spaces at location 1, at least 230 spaces at location 2, and at least 214 spaces at location 3.

EXAMPLE 7.3 (REVISITED)

We shall not give the details of the calculations, but shall simply state that the unique probability vector solution of the linear system $(I - P)\mathbf{q} = \mathbf{0}$ is

$$\mathbf{q} = \begin{bmatrix} 3/28 \\ 3/28 \\ 3/28 \\ 5/28 \\ 4/28 \\ 3/28 \\ 4/28 \\ 3/28 \end{bmatrix} = \begin{bmatrix} .1071... \\ .1071... \\ .1071... \\ .1785... \\ .1428... \\ .1071... \\ .1428... \\ .1071... \end{bmatrix}.$$

The entries in this vector indicate the proportion of time the officer spends at each intersection over the long term. Thus, if the objective is for him to spend the same proportion of time at each intersection, then the strategy of random movement with equal probabilities from one intersection to another is not a good one. (See Exercise 7.5.)

EXERCISES

7.1 For the transition matrix $P = \begin{bmatrix} .4 & .5 \\ .6 & .5 \end{bmatrix}$,

(a) calculate $x^{(n)}$ for $n = 1,2,3,4,5$ if $x^{(0)} = \begin{bmatrix} 1 \\ 0 \end{bmatrix}$.

(b) state why P is regular and find its steady-state vector.

7.2 For the transition matrix

$$P = \begin{bmatrix} .2 & .1 & .7 \\ .6 & .4 & .2 \\ .2 & .5 & .1 \end{bmatrix},$$

(a) calculate $x^{(1)}$, $x^{(2)}$, and $x^{(3)}$ to three decimal places if

$$x^{(0)} = \begin{bmatrix} 0 \\ 0 \\ 1 \end{bmatrix}.$$

(b) state why P is regular and find its steady-state vector.

7.3 Find the steady-state vectors of the following regular transition matrices:

(a) $\begin{bmatrix} 1/3 & 3/4 \\ 2/3 & 1/4 \end{bmatrix}$ (b) $\begin{bmatrix} .81 & .26 \\ .19 & .74 \end{bmatrix}$ (c) $\begin{bmatrix} 1/3 & 1/2 & 0 \\ 1/3 & 0 & 1/4 \\ 1/3 & 1/2 & 3/4 \end{bmatrix}$

7.4 Let P be the transition matrix

$$\begin{bmatrix} 1/2 & 0 \\ 1/2 & 1 \end{bmatrix}.$$

(a) Show that P is not regular.
(b) Show that as n increases, $P^n x^{(0)}$ approaches $\begin{bmatrix} 0 \\ 1 \end{bmatrix}$ for any initial state vector $x^{(0)}$.

(c) What conclusion of Theorem 7.3 is not valid for the steady-state vector of this transition matrix?

7.5 Verify that if P is a $k \times k$ regular transition matrix all of whose row sums are equal to one, then the entries of its steady-state vector are all equal to $1/k$.

7.6 Show that the transition matrix

$$P = \begin{bmatrix} 0 & 1/2 & 1/2 \\ 1/2 & 1/2 & 0 \\ 1/2 & 0 & 1/2 \end{bmatrix}$$

is regular, and use Exercise 7.5 to find its steady-state vector.

7.7 John is either happy or sad. If he is happy one day, he is happy the next day four times out of five. If he is sad one day, he is sad the next day one time out of three. Over the long term, what are the chances that John is happy on any given day?

7.8 A country is divided into three demographic regions. It is found that each year 5% of the residents of region 1 move to region 2 and 5% move to region 3. Of the residents of region 2, 15% move to region 1 and 10% move to region 3. And of the residents of region 3, 10% move to region 1 and 5% move to region 2. What percentage of the population resides in each of the three regions after a long period of time?

Leontief Economic Models

*Two linear models for economic systems are
discussed. Some results about nonnegative
matrices are applied to determine equilibrium price structures and outputs necessary
to satisfy demand.*

PREREQUISITES: Linear systems
Matrices

INTRODUCTION

Matrix theory has been very successful in describing the interrelations between prices, outputs, and demands in an economic
system. In this chapter, we discuss some simple models based on the
ideas of the Nobel-laureate Wassily Leontief. Two different, but
related, models are discussed: the closed, or input-output, model;
and the open, or production, model. In each, we are given certain
economic parameters which describe the interrelations between the
"industries" in the economy under consideration. Using matrix theory, we then evaluate certain other parameters, such as prices or
output levels, in order to satisfy a desired economic objective. We
begin, in the next section, with the closed model.

LEONTIEF CLOSED (INPUT-OUTPUT) MODEL

We begin with a simple example, and then proceed to the general theory of the model.

EXAMPLE 8.1
Three homeowners — a carpenter, an electrician, and a plumber — mutually agree to make repairs in their three homes. They agree to work a total of ten days each according to the following schedule:

	Work performed by		
	carpenter	electrician	plumber
Days of work in home of carpenter	2	1	6
Days of work in home of electrician	4	5	1
Days of work in home of plumber	4	4	3

For tax purposes, they must report and pay each other a reasonable daily wage, even for the work each does on his own home. Their normal daily wages are in the $60 - $80 range, but they agree to adjust their respective daily wages so that each homeowner will come out even; that is, so that the total amount paid out by each is the same as the total amount each receives. Let us set

p_1 = daily wage of carpenter

p_2 = daily wage of electrician

p_3 = daily wage of plumber.

To satisfy the "equilibrium" condition that each homeowner comes out even, we require that

total expenditures = total income

for each of the homeowners for the ten day period. For example, the carpenter pays a total of $2p_1 + p_2 + 6p_3$ for the repairs in his own home, and receives a total income of $10p_1$ for the repairs that he performs on all three homes. Equating these two expressions then gives the first of the following three equations:

$$2p_1 + p_2 + 6p_3 = 10p_1$$
$$4p_1 + 5p_2 + p_3 = 10p_2$$
$$4p_1 + 4p_2 + 3p_3 = 10p_3 \, .$$

The remaining two equations are the equilibrium equations for the electrician and the plumber. Dividing these equations by ten and rewriting them in matrix form yields

$$\begin{bmatrix} .2 & .1 & .6 \\ .4 & .5 & .1 \\ .4 & .4 & .3 \end{bmatrix} \begin{bmatrix} p_1 \\ p_2 \\ p_3 \end{bmatrix} = \begin{bmatrix} p_1 \\ p_2 \\ p_3 \end{bmatrix}. \qquad (8.1)$$

Equation (8.1) can be rewritten as a homogeneous system by subtracting the lefthand side from the righthand side to obtain

$$\begin{bmatrix} .8 & -.1 & -.6 \\ -.4 & .5 & -.1 \\ -.4 & -.4 & .7 \end{bmatrix} \begin{bmatrix} p_1 \\ p_2 \\ p_3 \end{bmatrix} = \begin{bmatrix} 0 \\ 0 \\ 0 \end{bmatrix}.$$

The solution of this homogeneous system is easily found to be (verify)

$$\begin{bmatrix} p_1 \\ p_2 \\ p_3 \end{bmatrix} = s \begin{bmatrix} 31 \\ 32 \\ 36 \end{bmatrix},$$

where s is an arbitrary constant. This constant is a scale factor which the homeowners may choose for their convenience. For example, they may set $s = 2$ so that the corresponding daily wages — $62, $64 and $72 — fall in the $60 - $80 range.

◆━◆━◆

This example illustrates the salient features of the Leontief input-output model of a closed economy. In the basic equation (8.1), each column sum of the coefficient matrix is one, corresponding to the fact that each of the homeowners' "output" of labor is completely distributed among these same homeowners in the proportions given by the entries in the column. Our problem was to determine suitable "prices" for these outputs to put the system in equilibrium; that is so that each homeowner's total expenditures equal his total income.

In the general model, we have an economic system consisting of a finite number of "industries", which we number as industries 1, 2, ..., k. Over some fixed period of time, each industry produces an "output" of some good or service which is completely utilized in a predetermined manner by the k industries. An important problem is to find suitable "prices" to be charged for these k outputs so that for each industry total expenditures equal total income. Such a price structure represents an equilibrium position for the economy.

For the fixed time period in question, let us set

p_i = price charged by the ith industry for its total output,

e_{ij} = fraction of the total output of the jth industry purchased by the ith industry.

for $i, j = 1, 2, \ldots, k$. By definition, we have

(i) $p_i \geq 0$, $\qquad\qquad\qquad\qquad$ $i = 1, 2, \ldots, k$.

(ii) $e_{ij} \geq 0$, $\qquad\qquad\qquad\quad$ $i, j = 1, 2, \ldots, k$.

(iii) $e_{1j} + e_{2j} + \cdots + e_{kj} = 1$, $\quad$ $j = 1, 2, \ldots, k$.

With these quantities, we form the *price vector*

$$p = \begin{bmatrix} p_1 \\ p_2 \\ \vdots \\ p_k \end{bmatrix}$$

and the *exchange* or *input-output matrix*

$$E = \begin{bmatrix} e_{11} & e_{12} & \cdots & e_{1k} \\ e_{21} & e_{22} & \cdots & e_{2k} \\ \vdots & \vdots & & \vdots \\ e_{k1} & e_{k2} & \cdots & e_{kk} \end{bmatrix}.$$

Condition (iii) above expresses the fact that all of the column sums of the exchange matrix are one.

As in the example, in order that the expenditures of each industry be equal to its income, the following matrix equation must be satisfied (see (8.1)):

$$Ep = p \tag{8.2}$$

or

$$(I - E)p = 0. \tag{8.3}$$

Equation (8.3) is a homogeneous linear system for the price vector p. It will have a nontrivial solution if and only if the determinant of its coefficient matrix $I - E$ is zero. In Exercise 8.6 we ask the reader to show that this is the case for any exchange matrix E. Thus, (8.3) always has nontrivial solutions for the price vector p.

   Actually, for our economic model to make sense we need more than just the fact that (8.3) has nontrivial solutions for p. We also need that the prices p_i of the k outputs are nonnegative numbers. We express this condition as $p \geq 0$. (In general, if A is any vector or matrix, the notation $A \geq 0$ means that every entry of A is nonnegative, and the notation $A > 0$ means that every entry of A is positive. Similarly, $A \geq B$ denotes $A - B \geq 0$, and $A > B$ denotes $A - B > 0$.) To show that (8.3) has a nontrivial solution for which $p \geq 0$ is a bit more difficult than showing merely that some nontrivial solution exists. But it is true, and we state this fact without proof in the following theorem:

THEOREM 8.1 *If E is an exchange matrix, then $Ep = p$ always has a nontrivial solution p whose entries are nonnegative.*

Let us consider a few simple examples of this theorem.

EXAMPLE 8.2

   Let

$$E = \begin{bmatrix} 1/2 & 0 \\ 1/2 & 1 \end{bmatrix}.$$

Then $(I - E)p = 0$ is

$$\begin{bmatrix} 1/2 & 0 \\ -1/2 & 0 \end{bmatrix} \begin{bmatrix} p_1 \\ p_2 \end{bmatrix} = \begin{bmatrix} 0 \\ 0 \end{bmatrix},$$

which has the general solution

$$p = s \begin{bmatrix} 0 \\ 1 \end{bmatrix},$$

where s is an arbitrary constant. We then have nontrivial solutions $p \geq 0$ for any $s > 0$.

EXAMPLE 8.3

   Let

$$E = \begin{bmatrix} 1 & 0 \\ 0 & 1 \end{bmatrix}.$$

Then $(I - E)\mathbf{p} = 0$ has the general solution

$$\mathbf{p} = s\begin{bmatrix} 1 \\ 0 \end{bmatrix} + t\begin{bmatrix} 0 \\ 1 \end{bmatrix}.$$

where s and t are independent arbitrary constants. Nontrivial solutions $\mathbf{p} \geq 0$ then result from any $s \geq 0$ and $t \geq 0$, not both zero.

◆━◆━◆

Example 8.2 indicates that in some situations one of the prices must be zero in order to satisfy the equilibrium condition. Example 8.3 indicates that there may be several linearly independent price structures available. Neither of these situations describes a truly interdependent economic structure. The following theorem gives sufficient conditions for both cases to be excluded.

THEOREM 8.2 *Let E be an exchange matrix such that for some positive integer m all of the entries of E^m are positive. Then there is exactly one linearly independent solution of $(I - E)\mathbf{p} = 0$, and it may be chosen so that all of its entries are positive.*

We shall not give a proof of this theorem. The reader who has read Chapter 7 on Markov Chains may observe that this theorem is essentially the same as Theorem 7.3. What we are calling exchange matrices in this chapter were called transition or Markov matrices in Chapter 7.

EXAMPLE 8.4

The exchange matrix in Example 8.1 was

$$E = \begin{bmatrix} .2 & .1 & .6 \\ .4 & .5 & .1 \\ .4 & .4 & .3 \end{bmatrix}.$$

Since $E > 0$, the condition $E^m > 0$ in Theorem 8.2 is satisfied for $m = 1$. Consequently, we are guaranteed that there is exactly one linearly independent solution of $(I - E)\mathbf{p} = 0$, and it may be chosen so that $\mathbf{p} > 0$. In that example, we found that

$$\mathbf{p} = \begin{bmatrix} 31 \\ 32 \\ 36 \end{bmatrix}$$

is such a solution.

LEONTIEF OPEN (PRODUCTION) MODEL

In contrast with the closed model in which the outputs of k industries are distributed only among themselves, the open model attempts to satisfy an outside demand for the outputs. Portions of these outputs may still be distributed among the industries themselves, to keep them operating, but there is to be some excess, some net production, with which to satisfy the outside demand. In the closed model, the outputs of the industries were fixed, and our objective was to determine prices for these outputs so that the equilibrium condition, that expenditures equal incomes, was satisfied. In the open model, it is the prices which are fixed, and our objective will be to determine levels of the outputs of the industries needed to satisfy the outside demand. We shall measure the levels of the outputs in terms of their economic values using the fixed prices. To be precise, over some fixed period of time let

x_i = monetary value of the total output of the ith industry,

d_i = monetary value of the output of the ith industry needed to satisfy the outside demand,

c_{ij} = monetary value of the output of the ith industry needed by the jth industry to produce one unit of monetary value of its own output.

With these quantities, we define the *production vector*

$$\mathbf{x} = \begin{bmatrix} x_1 \\ x_2 \\ \vdots \\ x_k \end{bmatrix},$$

the *demand vector*

$$\mathbf{d} = \begin{bmatrix} d_1 \\ d_2 \\ \vdots \\ d_k \end{bmatrix},$$

and the *consumption matrix*

$$C = \begin{bmatrix} c_{11} & c_{12} & \cdots & c_{1k} \\ c_{21} & c_{22} & \cdots & c_{2k} \\ \vdots & \vdots & & \vdots \\ c_{k1} & c_{k2} & \cdots & c_{kk} \end{bmatrix}.$$

By their nature, we have that

$$\mathbf{x} \geq 0, \quad \mathbf{d} \geq 0, \quad \text{and} \quad C \geq 0.$$

From the definition of c_{ij} and x_j, it can be seen that the quantity

$$c_{i1}x_1 + c_{i2}x_2 + \cdots + c_{ik}x_k$$

is the value of the output of the ith industry needed by all k industries to produce a total output specified by the production vector $\mathbf{x}$. Since this quantity is simply the ith entry of the column vector $C\mathbf{x}$, we can further say that the ith entry of the column vector

$$\mathbf{x} - C\mathbf{x}$$

is the value of the excess output of the ith industry available to satisfy the outside demand. The value of the outside demand for the output of the ith industry is the ith entry of the demand vector $\mathbf{d}$. Consequently, we are led to the following equation

$$\mathbf{x} - C\mathbf{x} = \mathbf{d}$$

or

$$(I - C)\mathbf{x} = \mathbf{d} \tag{8.4}$$

for the demand to be exactly met, without any surpluses or shortages. Thus, given C and $\mathbf{d}$, our objective is to find a production vector $\mathbf{x} \geq 0$ which satisfies Equation (8.4).

EXAMPLE 5.5

A town has three main industries: a coal-mining operation, an electric power generating plant, and a local railroad. To mine $1 of coal, the mining operation must purchase $.25 of electricity to run its equipment and $.25 of transportation for its shipping needs. To produce $1 of electricity, the generating plant requires $.65 of coal for fuel, $.05 of its own electricity to run auxilliary equipment, and $.05 of transportation. To provide $1 of transportation, the railroad requires $.55 of coal for fuel, and $.10 of electricity for its auxilliary equipment. In a certain week, the

coal-mining operation receives orders for $50,000 of coal from out-
side the town, and the generating plant receives orders for $25,000
of electricity from outside. There is no outside demand for the
local railroad. How much must each of the three industries produce
in that week to exactly satisfy their own demand and the outside demand?

SOLUTION For the one week period, let

x_1 = value of total output of coal-mining operation,

x_2 = value of total output of power generating plant,

x_3 = value of total output of local railroad.

From the information supplied, the consumption matrix of the system
is

$$C = \begin{bmatrix} 0 & .65 & .55 \\ .25 & .05 & .10 \\ .25 & .05 & 0 \end{bmatrix}.$$

The linear system $(I - C)x = d$ is then

$$\begin{bmatrix} 1.00 & -.65 & -.55 \\ -.25 & .95 & -.10 \\ -.25 & -.05 & 1.00 \end{bmatrix} \begin{bmatrix} x_1 \\ x_2 \\ x_3 \end{bmatrix} = \begin{bmatrix} 50,000 \\ 25,000 \\ 0 \end{bmatrix}.$$

The coefficient matrix on the left is invertible, and the solution
is given by

$$x = (I - C)^{-1}d = \frac{1}{503} \begin{bmatrix} 756 & 542 & 470 \\ 220 & 690 & 190 \\ 200 & 170 & 630 \end{bmatrix} \begin{bmatrix} 50,000 \\ 25,000 \\ 0 \end{bmatrix} = \begin{bmatrix} 102,087 \\ 56,163 \\ 28,330 \end{bmatrix}.$$

Thus, the total output of the coal-mining operation should be
$102,087, the total output of the power generating plant should be
$56,163, and the total output of the railroad should be $28,330.

◆━◆━◆

Let us reconsider Eq. (8.4):

$$(I - C)x = d.$$

If the square matrix $I - C$ is invertible, we may write

$$x = (I - C)^{-1}d. \qquad\qquad (8.5)$$

In addition, if the matrix $(I - C)^{-1}$ has only nonnegative entries, then we are guaranteed that for any $d \geq 0$, Eq. (8.5) determines a unique nonnegative solution for x. This is a particularly desirable situation since it means that any outside demand can be met. The terminology used to describe this case is given in the following definition:

> **DEFINITION 8.1** *A consumption matrix C is said to be* **productive** *if $(I - C)^{-1}$ exists and $(I - C)^{-1} \geq 0$.*

We shall now consider some simple criteria which will guarantee that a consumption matrix is productive. The first is given in the following theorem.

> **THEOREM 8.3** *A consumption matrix C is productive if and only if there is some production vector $x \geq 0$ such that $x > Cx$.*

(The proof is outlined in Exercise 8.8.) The condition $x > Cx$ means that there is some production schedule possible such that each industry produces more than it consumes.

Theorem 8.3 has two interesting corollaries. Suppose all of the row sums of C are less than one. Then if

$$x = \begin{bmatrix} 1 \\ 1 \\ \vdots \\ 1 \end{bmatrix},$$

Cx is a column vector whose entries are these row sums. Thus $x > Cx$, and the condition of Theorem 8.3 is satisfied. We state this as

> **COROLLARY 8.1** *A consumption matrix is productive if each of its row sums is less than one.*

As we ask the reader to show in Exercise 8.7, this corollary leads to

> **COROLLARY 8.2** *A consumption matrix is productive if each of its column sums is less than one.*

Recalling the definition of the entries of the consumption matrix C, we see that the jth column sum of C is the total value of the outputs of all k industries needed to produce one unit of value of output of the jth industry. The jth industry is thus said to be *profitable* if that jth column sum is less than one. In other words, Corollary 8.2 says that a consumption matrix is productive if all k industries in the economic system are profitable.

EXAMPLE 8.6 The consumption matrix in Example 8.5 was

$$C = \begin{bmatrix} 0 & .65 & .55 \\ .25 & .05 & .10 \\ .25 & .05 & 0 \end{bmatrix}.$$

All three column sums in this matrix are less than one, and so all three industries are profitable. Consequently, by Corollary 8.2, the consumption matrix C is productive. This can also be seen in the calculations in Example 8.5 since $(I - C)^{-1}$ is nonnegative.

EXERCISES

8.1 For the following exchange matrices, find nonnegative price vectors which satisfy the equilibrium condition (8.3).

(a) $\begin{bmatrix} 1/2 & 1/3 \\ 1/2 & 2/3 \end{bmatrix}$
(b) $\begin{bmatrix} 1/2 & 0 & 1/2 \\ 1/3 & 0 & 1/2 \\ 1/6 & 1 & 0 \end{bmatrix}$
(c) $\begin{bmatrix} .35 & .50 & .30 \\ .25 & .20 & .30 \\ .40 & .30 & .40 \end{bmatrix}$

8.2 Using Theorem 8.3 and its corollaries, show that each of the following consumption matrices is productive.

(a) $\begin{bmatrix} .8 & .1 \\ .3 & .6 \end{bmatrix}$
(b) $\begin{bmatrix} .70 & .30 & .25 \\ .20 & .40 & .25 \\ .05 & .15 & .25 \end{bmatrix}$
(c) $\begin{bmatrix} .7 & .3 & .2 \\ .1 & .4 & .3 \\ .2 & .4 & .1 \end{bmatrix}$

8.3 Using Theorem 8.2, show that there is only one linearly independent price vector for the closed economic system with exchange matrix

$$E = \begin{bmatrix} 0 & .2 & .5 \\ 1 & .2 & .5 \\ 0 & .6 & 0 \end{bmatrix}.$$

8.4 Three neighbors have backyard vegetable gardens. Neighbor A grows tomatoes, neighbor B grows corn, and neighbor C grows lettuce. They agree to divide their crops among themselves as follows: A gets 1/2 of the tomatoes, 1/3 of the corn, and 1/4 of the lettuce. B gets 1/3 of the tomatoes, 1/3 of the corn, and 1/4 of the lettuce. C gets 1/6 of the tomatoes, 1/3 of the corn, and 1/2 of the lettuce. What prices should the neighbors assign to their respective crops if the equilibrium condition of a closed economy is to be satisfied, and if the lowest-priced crop is to have a price of $100?

8.5 Three engineers — a civil engineer (CE), an electrical engineer (EE), and a mechanical engineer (ME) — each have a consulting firm. The consulting they do is of a multidisciplinary nature, and so they buy a portion of each others' services. For each $1 of consulting the CE does, he buys $.10 of the EE's services and $.30 of the ME's services. For each $1 of consulting the EE does, he buys $.20 of the CE's services and $.40 of the ME's services. And for each $1 of consulting the ME does, he buys $.30 of the CE's services and $.40 of the EE's services. In a certain week, the CE receives outside consulting orders of $500, the EE receives outside consulting orders of $700, and the ME receives outside consulting orders of $600. What dollar amount of consulting does each engineer perform in that week?

8.6 Using the fact that the column sums of an exchange matrix E are all one, show that the column sums of $I - E$ are zero. From this, show that $I - E$ has zero determinant, and so $(I - E)p = 0$ has nontrivial solutions for p.

8.7 Show that Corollary 8.2 follows from Corollary 8.1. Hint: use the fact that $(A^t)^{-1} = (A^{-1})^t$ for any invertible matrix A.

8.8 Prove Theorem 8.3 as follows:

I. Prove the "only if" part of theorem; i.e. show that if C is a productive consumption matrix, then there is a vector $x \geq 0$ such that $x > Cx$.

II. Prove the "if" part of theorem as follows:

Step 1. Show that if there is a vector $x^* \geq 0$ such that $Cx^* < x^*$, then $x^* > 0$.

Step 2. Show that there is a number λ such that $0 < \lambda < 1$ and $Cx^* < \lambda x^*$.

Step 3. Show that $C^n x^* < \lambda^n x^*$ for $n = 1, 2, \ldots$.

Step 4. Show that $C^n \to 0$ as $n \to \infty$.

Step 5. By direct verification, show that

$$(I - C)(I + C + C^2 + \cdots + C^{n-1}) = I - C^n$$

for $n = 1, 2, \ldots$.

Step 6. By letting $n \to \infty$ in step 5, show that the matrix infinite sum

$$S = I + C + C^2 + \cdots$$

exists and that $(I - C)S = I$.

Step 7. Show that $S \geq 0$ and that $S = (I - C)^{-1}$.

Step 8. Conclude from Definition 5.1 that C is a productive consumption matrix.

9

Forest
Management

A matrix model for the management of a forest
whose trees are grouped into height classes
is presented. The optimal sustainable yield
of a periodic harvest is calculated when the
trees of different height classes can have
different economic values.

PREREQUISITES: Matrix multiplication and addition

INTRODUCTION

In this chapter, we shall introduce a simplified model for the sustainable harvesting of a forest whose trees are classified by height. The height of a tree is assumed to determine its economic value when it is cut down and sold. Initially, there is a distribution of trees of various heights. The forest is then allowed to grow for a certain period of time, after which some of the trees of various heights are harvested. But the trees left unharvested are to be of the same height configuration as the original forest, so that the harvest is sustainable. As we shall see, there are many such sustainable harvesting procedures. We want to find one for which the total economic value of all the trees removed is as large as possible. This determines the *optimal sustainable yield* of the forest, and is the largest yield which can be attained continually without depleting the forest.

THE MODEL

Suppose a harvester has a forest of Douglas fir trees which he wants to sell as Christmas trees year after year. Every December he cuts down some of the trees to be sold. For each tree he cuts down, he plants a seedling in its place. In this way, the total number of trees in the forest is always the same. (In this simplified model, we shall not take into account trees which die between harvests. We assume that every seedling planted survives and grows until it is harvested.)

In the market place, trees of different heights have different economic values. Suppose there are n different price classes corresponding to certain height intervals as follows:

Class	Value in Dollars	Height Interval
1 (seedlings)	none	$[0, h_1)$
2	p_2	$[h_1, h_2)$
3	p_3	$[h_2, h_3)$
$\vdots$	$\vdots$	$\vdots$
$n-1$	p_{n-1}	$[h_{n-2}, h_{n-1})$
n	p_n	$[h_{n-1}, \infty)$

Figure 9.1

The first class consists of seedlings with heights in the interval $[0, h_1)$, and these seedlings are of no economic value. The n-th class consists of trees with heights greater than or equal to h_{n-1}.

Let $x_i (i = 1, 2, \ldots, n)$ be the number of trees within the i-th class which remain after each harvest. We form a column vector with these numbers

$$\mathbf{x} = \begin{bmatrix} x_1 \\ x_2 \\ \vdots \\ x_n \end{bmatrix}$$

and call it the *nonharvest vector*. For a sustainable harvesting policy, the forest is to be returned after each harvest to the fixed configuration given by the nonharvest vector $\mathbf{x}$. Part of our problem will be to find those nonharvest vectors $\mathbf{x}$ for which sustainable harvesting is possible.

Since the total number of trees in the forest is fixed, let us set

$$x_1 + x_2 + \cdots + x_n = s \tag{9.1}$$

where s is predetermined by the amount of land available and the amount of space each tree requires. Referring to Fig. 9.2, we have the following situation: The forest configuration is given by the vector $\mathbf{x}$ after each harvest. Between harvests, the trees grow and produce a new forest configuration before each harvest. A certain number of trees are removed from each class at the harvest. Finally, a seedling is planted in place of each tree removed to return the forest again to the configuration $\mathbf{x}$.

Let us first consider the growth of the forest between harvests. During this period, a tree in the i-th class may grow and move up to a higher height class. Or its growth may be retarded for some reason, and it will remain in the same class. We consequently define the following growth parameters, g_i, for $i = 1, 2, \ldots, n-1$:

g_i = the fraction of trees in the i-th class that grow into the $(i + 1)$-st class during a growth period.

For simplicity, we shall assume that a tree can move at most one height class upward in one growth period. With this assumption we have

$1 - g_i$ = the fraction of trees in the i-th class that remain in the i-th class during a growth period.

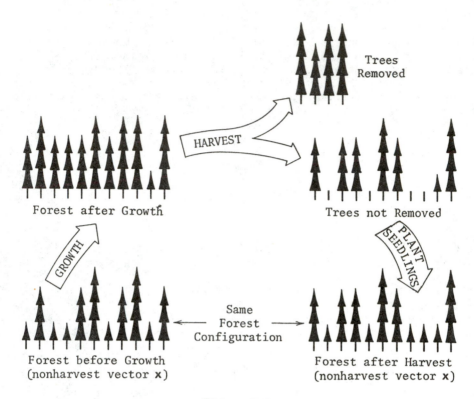

Figure 9.2

With these $n-1$ growth parameters, we form the following $n \times n$ *growth matrix*:

$$
G = \begin{bmatrix}
1-g_1 & 0 & 0 & \cdots & 0 \\
g_1 & 1-g_2 & 0 & \cdots & 0 \\
0 & g_2 & 1-g_3 & \cdots & 0 \\
\vdots & \vdots & \vdots & & \vdots \\
0 & 0 & 0 & \cdots 1-g_{n-1} & 0 \\
0 & 0 & 0 & \cdots & g_{n-1} & 1
\end{bmatrix}.
\tag{9.2}
$$

Since the entries of the vector x are the numbers of trees in the n classes before the growth period, the reader can verify that the entries of the vector

$$
G\mathbf{x} = \begin{bmatrix} (1-g_1)x_1 \\ g_1x_1 + (1-g_2)x_2 \\ g_2x_2 + (1-g_3)x_3 \\ \vdots \\ g_{n-2}x_{n-2} + (1-g_{n-1})x_{n-1} \\ g_{n-1}x_{n-1} + x_n \end{bmatrix} \tag{9.3}
$$

are the numbers of trees in the n classes after the growth period.

Suppose during the harvest we remove $y_i (i = 1, 2, \ldots, n)$ trees from the i-th class. We shall call the column vector

$$
\mathbf{y} = \begin{bmatrix} y_1 \\ y_2 \\ \vdots \\ y_n \end{bmatrix}
$$

the *harvest vector*. Thus, a total of

$$
y_1 + y_2 + \cdots + y_n
$$

trees are removed at each harvest. This is also the total number of trees added to the first class (the new seedlings) after each harvest. If we define the following $n \times n$ *replacement matrix*

$$
R = \begin{bmatrix} 1 & 1 & 1 & \cdots & 1 & 1 \\ 0 & 0 & 0 & \cdots & 0 & 0 \\ \vdots & \vdots & \vdots & & \vdots & \vdots \\ 0 & 0 & 0 & \cdots & 0 & 0 \end{bmatrix} \tag{9.4}
$$

then the column vector

$$
R\mathbf{y} = \begin{bmatrix} y_1 + y_2 + \cdots + y_n \\ 0 \\ 0 \\ \vdots \\ 0 \end{bmatrix} \tag{9.5}
$$

specifies the configuration of trees planted after each harvest.

At this point, we are ready to write the following equation which characterizes a sustainable harvesting policy:

$$\begin{pmatrix} \text{configuration} \\ \text{at end of} \\ \text{growth period} \end{pmatrix} - \Big(\text{harvest}\Big) + \begin{pmatrix} \text{new seedling} \\ \text{replacement} \end{pmatrix} = \begin{pmatrix} \text{configuration} \\ \text{at beginning of} \\ \text{growth period} \end{pmatrix}$$

or mathematically

$$Gx - y + Ry = x.$$

This equation can be rewritten as

$$(I - R)y = (G - I)x, \tag{9.6}$$

or, more fully,

$$\begin{bmatrix} 0 & -1 & -1 & \cdots & -1 & -1 \\ 0 & 1 & 0 & \cdots & 0 & 0 \\ 0 & 0 & 1 & \cdots & 0 & 0 \\ \vdots & \vdots & \vdots & & \vdots & \vdots \\ 0 & 0 & 0 & \cdots & 1 & 0 \\ 0 & 0 & 0 & \cdots & 0 & 1 \end{bmatrix} \begin{bmatrix} y_1 \\ y_2 \\ y_3 \\ \vdots \\ y_{n-1} \\ y_n \end{bmatrix} = \begin{bmatrix} -g_1 & 0 & 0 & \cdots & 0 & 0 \\ g_1 & -g_2 & 0 & \cdots & 0 & 0 \\ 0 & g_2 & -g_3 & \cdots & 0 & 0 \\ \vdots & \vdots & \vdots & & \vdots & \vdots \\ 0 & 0 & 0 & \cdots & -g_{n-1} & 0 \\ 0 & 0 & 0 & \cdots & g_{n-1} & 0 \end{bmatrix} \begin{bmatrix} x_1 \\ x_2 \\ x_3 \\ \vdots \\ x_{n-1} \\ x_n \end{bmatrix}$$

We shall refer to Eq. (9.6) as the *sustainable harvesting condition*. Any vectors x and y with nonnegative entries, and such that $x_1 + x_2 + \cdots + x_n = s$, which satisfy this matrix equation determine a sustainable harvesting policy for the forest. Let us note that if $y_1 > 0$, then the harvester is removing seedlings, of no economic value, and then replacing them with new seedlings. Since there is no point in doing this, we shall assume

$$y_1 = 0. \tag{9.7}$$

With this assumption, it may be verified that (9.6) is the matrix form of the following set of equations:

$$y_2 + y_3 + \cdots + y_n = g_1 x_1$$
$$y_2 = g_1 x_1 - g_2 x_2$$
$$y_3 = g_2 x_2 - g_3 x_3$$
$$\vdots \qquad\qquad\qquad (9.8)$$
$$y_{n-1} = g_{n-2} x_{n-2} - g_{n-1} x_{n-1}$$
$$y_n = g_{n-1} x_{n-1}.$$

Notice that the first of Eqs. (9.8) is the sum of the remaining $n-1$ equations.

Since we must have $y_i \geq 0$ for $i = 2, 3, \ldots, n$, Eqs. (9.8) require that

$$g_1 x_1 \geq g_2 x_2 \geq \cdots \geq g_{n-1} x_{n-1} \geq 0. \qquad (9.9)$$

Conversely, if x is a column vector with nonnegative entries which satisfy Eq. (9.9), then Eqs. (9.7) and (9.8) define a column vector y with nonnegative entries. Furthermore, x and y then satisfy the sustainable harvesting condition (9.6). In other words, a necessary and sufficient condition that a nonnegative column vector x determine a forest configuration which is capable of sustainable harvesting is that its entries satisfy (9.9).

OPTIMAL SUSTAINABLE YIELD

Since we remove y_i trees from the i-th class ($i = 2, 3, \ldots, n$) and since each tree in the i-th class has an economic value of p_i, the total yield of the harvest, Yld, is given by

$$Yld = p_2 y_2 + p_3 y_3 + \cdots + p_n y_n. \qquad (9.10)$$

Using (9.8), we may substitute for the y_i's in (9.10) to obtain

$$Yld = p_2 g_1 x_1 + (p_3 - p_2) g_2 x_2 + \cdots + (p_n - p_{n-1}) g_{n-1} x_{n-1}. \qquad (9.11)$$

Combining Eqs. (9.11), (9.1), and (9.9), we may now state the problem of maximizing the yield of the forest over all possible sustainable harvesting policies as

Problem *Find nonnegative $x_1, x_2, \ldots, x_n$ which maximize*

$$Yld = p_2 g_1 x_1 + (p_3 - p_2) g_2 x_2 + \cdots + (p_n - p_{n-1}) g_{n-1} x_{n-1}$$

subject to

$$x_1 + x_2 + \cdots + x_n = s$$

and

$$g_1 x_1 \geq g_2 x_2 \geq \cdots \geq g_{n-1} x_{n-1} \geq 0.$$

As formulated above, this problem belongs to the field of Linear Programming. (See Chapter 20 for an introduction to this field.) However, we shall only need the following result from Linear Programming theory applied to this problem:

The optimal sustainable yield is achieved by harvesting all of the trees from one particular height class and none of the trees from any other height class.

Below, we prove this result without Linear Programming theory by actually exhibiting such a sustainable harvesting policy.

Let us first set

$$Yld_k = \text{yield obtained by harvesting all of the } k\text{-th}$$
class and none of the other classes.

The largest value of Yld_k for $k = 2, 3, \ldots, n$ will then be the optimal sustainable yield, and the corresponding value of k will be the class which should be completely harvested to attain the optimal sustainable yield. Since no class but the k-th is harvested, we have

$$y_2 = y_3 = \cdots = y_{k-1} = y_{k+1} = \cdots = y_n = 0. \tag{9.12}$$

In addition, since all of the k-th class is harvested, no trees are left unharvested in the k-th class, and no trees are ever present in the height classes above the k-th class. Thus,

$$x_k = x_{k+1} = \cdots = x_n = 0. \tag{9.13}$$

Substitution of (9.12) and (9.13) into the sustainable harvesting condition (9.8) then gives

$$
\begin{aligned}
y_k &= g_1 x_1 \\
0 &= g_1 x_1 - g_2 x_2 \\
0 &= g_2 x_2 - g_3 x_3 \\
&\vdots \\
0 &= g_{k-2} x_{k-2} - g_{k-1} x_{k-1} \\
y_k &= g_{k-1} x_{k-1}.
\end{aligned}
\tag{9.14}
$$

Equations (9.14) may also be written as

$$y_k = g_1 x_1 = g_2 x_2 = \cdots = g_{k-1} x_{k-1}, \tag{9.15}$$

from which it follows that

$$
\begin{aligned}
x_2 &= g_1 x_1 / g_2 \\
x_3 &= g_1 x_1 / g_3 \\
&\vdots \\
x_{k-1} &= g_1 x_1 / g_{k-1}.
\end{aligned}
\tag{9.16}
$$

If we substitute Eqs. (9.13) and (9.16) into (9.1),

$$x_1 + x_2 + \cdots + x_n = s,$$

we may solve for x_1 and obtain

$$x_1 = \frac{s}{1 + \dfrac{g_1}{g_2} + \dfrac{g_1}{g_3} + \cdots + \dfrac{g_1}{g_{k-1}}}. \tag{9.17}$$

For the yield Yld_k, we combine Eqs. (9.10), (9.12), (9.15) and (9.17) to obtain

$$
\begin{aligned}
Yld_k &= p_2 y_2 + p_3 y_3 + \cdots + p_n y_n \\
&= p_k y_k \\
&= p_k g_1 x_1 \\
&= \frac{p_k s}{\dfrac{1}{g_1} + \dfrac{1}{g_2} + \cdots + \dfrac{1}{g_{k-1}}}
\end{aligned}
\tag{9.18}
$$

Equation (9.18) determines Yld_k in terms of the known growth and economic parameters for any $k = 2, 3, \ldots, n$. Thus, the optimal sustainable yield is found as follows:

THEOREM 9.1 *The optimal sustainable yield is the largest value of*

$$
\frac{p_k s}{\dfrac{1}{g_1} + \dfrac{1}{g_2} + \cdots + \dfrac{1}{g_{k-1}}}
$$

for $k = 2, 3, \ldots, n$. The corresponding value of k is the number of the class which is completely harvested.

In Exercise (9.4), we ask the reader to show that the nonharvest vector x for the optimal sustainable yield is

$$
x = \frac{s}{\dfrac{1}{g_1} + \dfrac{1}{g_2} + \cdots + \dfrac{1}{g_{k-1}}}
\begin{bmatrix}
1/g_1 \\
1/g_2 \\
\vdots \\
1/g_{k-1} \\
0 \\
0 \\
\vdots \\
0
\end{bmatrix}.
\tag{9.19}
$$

Theorem 9.1 implies that it is not necessarily the highest-priced class of trees which should be totally cropped. The growth parameters, g_i, must also be taken into account to determine the optimal sustainable yield.

EXAMPLE 9.1 For a Scots Pine forest in Scotland with a growth period of six years, the following growth matrix was found (M. B. Usher, "A Matrix Approach to the Management of Renewable Resources, with Special Reference to Selection Forests," *Journal of Applied Ecology*, Vol. 3, 1966, pages 355 - 367):

$$G = \begin{bmatrix} .72 & 0 & 0 & 0 & 0 & 0 \\ .28 & .69 & 0 & 0 & 0 & 0 \\ 0 & .31 & .75 & 0 & 0 & 0 \\ 0 & 0 & .25 & .77 & 0 & 0 \\ 0 & 0 & 0 & .23 & .63 & 0 \\ 0 & 0 & 0 & 0 & .37 & 1.00 \end{bmatrix}.$$

Suppose the prices of trees in the five tallest height classes are

$$p_2 = \$50, \quad p_3 = \$100, \quad p_4 = \$150, \quad p_5 = \$200, \quad p_6 = \$250.$$

Which class should be completely harvested to obtain the optimal sustainable yield, and what is the yield?

SOLUTION From the matrix G we have that

$$g_1 = .28, \quad g_2 = .31, \quad g_3 = .25, \quad g_4 = .23, \quad g_5 = .37 .$$

Equation (9.18) then gives

$$Yld_2 = 50s/(.28^{-1}) = 14.0s$$

$$Yld_3 = 100s/(.28^{-1} + .31^{-1}) = 14.7s$$

$$Yld_4 = 150s/(.28^{-1} + .31^{-1} + .25^{-1}) = 13.9s$$

$$Yld_5 = 200s/(.28^{-1} + .31^{-1} + .25^{-1} + .23^{-1}) = 13.2s$$

$$Yld_6 = 250s/(.28^{-1} + .31^{-1} + .25^{-1} + .23^{-1} + .37^{-1}) = 14.0s.$$

We see that Yld_3 is the largest of these five quantities, so from Theorem 9.1 the third class should be completely harvested every six years to maximize the sustainable yield. The corresponding optimal sustainable yield is $\$14.7s$, where s is the total number of trees in the forest.

Exercises

9.1 A certain forest is divided into three height classes and has a growth matrix between harvests given by

$$G = \begin{bmatrix} 1/2 & 0 & 0 \\ 1/2 & 1/3 & 0 \\ 0 & 2/3 & 1 \end{bmatrix}.$$

If the price of trees in the second class is \$30 and the price of trees in the third class is \$50, which class should be completely harvested to attain the optimal sustainable yield? What is the optimal yield if there are 1,000 trees in the forest?

9.2 In Example 9.1, to what level must the price of trees in the fifth class rise so that class is the one to completely harvest in order to attain the optimal sustainable yield?

9.3 In Example 9.1, what must the ratio of the prices $p_2 : p_3 : p_4 : p_5 : p_6$ be in order that the yields Yld_k, $k = 2, 3, 4, 5, 6$, all be the same? (In this case, any sustainable harvesting policy will produce the same optimal sustainable yield.)

9.4 Derive Eq. (9.19) for the nonharvest vector x corresponding to the optimal sustainable harvesting policy described in Theorem 9.1.

9.5 For the optimal sustainable harvesting policy described in Theorem 9.1, how many trees are removed from the forest during each harvest?

9.6 If all of the growth parameters $g_1, g_2, \ldots, g_{n-1}$ in the growth matrix G are equal, what should the ratio of the prices $p_2 : p_3 : \ldots : p_n$ be in order that any sustainable harvesting policy be an optimal sustainable harvesting policy? (See Exercise 9.3.)

10

Computer Graphics

We assume that a view of a three-dimensional object is displayed on a video screen. Using matrix algebra, new views of the object are generated by rotation, translation, and scaling.

PREREQUISITES: Analytic geometry
                       Matrix algebra

INTRODUCTION

Let us suppose we want to visualize a three-dimensional object by displaying various views of it on a video screen. The object we have in mind to display is to be determined by a finite number of straight line segments. As an example, consider the truncated right pyramid with hexagonal base illustrated in Fig. 10.1. We first introduce an xyz-coordinate system in which to embed the object. As in Fig. 10.1, we orient the coordinate system so that its origin is at the center of the video screen and the xy-plane coincides with the plane of the screen. Consequently, an observer will see only the projection of the view of the three-dimensional object onto the two-dimensional xy-plane.

In the xyz-coordinate system, the endpoints $P_1, P_2, \ldots, P_n$ of the straight line segments which determine the view of the object will have certain coordinates, say

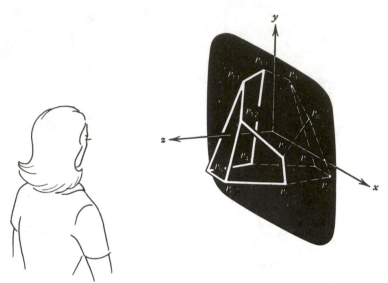

Figure 10.1

$$(x_1, y_1, z_1) \;, \quad (x_2, y_2, z_2) \;, \quad \ldots \;, \quad (x_n, y_n, z_n)$$

These coordinates, together with a specification of which pairs are to be connected by straight line segments, are to be stored in the memory of the video display system. For example, the truncated pyramid oriented as in Fig. 10.1 has twelve vertices with the following coordinates (the screen is four units wide by three units high):

P_1: (1.000,-.800,.000), P_2: (.500,-.800,-.866), P_3: (-.500,-.800,-.866)

P_4: (-1.000,-.800,.000), P_5: (-.500,-.800,.866), P_6: (.500,-.800,.866)

P_7: (.840,-.400,.000), P_8: (.315,.125,-.546), P_9: (-.210,.650,-.364)

P_{10}:(-.360,.800,.000), P_{11}:(-.210,.650,.364), P_{12}:(.315,.125,.546)

These 12 vertices are connected pairwise by 18 straight line segments as follows, where $P_i \leftrightarrow P_j$ denotes that point P_i is connected to point P_j:

$P_1 \leftrightarrow P_2$, $P_2 \leftrightarrow P_3$, $P_3 \leftrightarrow P_4$, $P_4 \leftrightarrow P_5$, $P_5 \leftrightarrow P_6$, $P_6 \leftrightarrow P_1$,

$P_7 \leftrightarrow P_8$, $P_8 \leftrightarrow P_9$, $P_9 \leftrightarrow P_{10}$, $P_{10} \leftrightarrow P_{11}$, $P_{11} \leftrightarrow P_{12}$, $P_{12} \leftrightarrow P_7$,

$P_1 \leftrightarrow P_7$, $P_2 \leftrightarrow P_8$, $P_3 \leftrightarrow P_9$, $P_4 \leftrightarrow P_{10}$, $P_5 \leftrightarrow P_{11}$, $P_6 \leftrightarrow P_{12}$.

In View 1 these 18 straight
segments are shown as they
would appear on the video
screen. It should be no-
ticed that only the x- and
y-coordinates of the vertices are needed by the
video display system to
draw the view since only
the projection of the ob-
ject onto the xy-plane is
displayed. However, we
must keep track of the
z-coordinates to carry out
certain transformations
discussed later.

View 1

We now show how to form new
views of the object produced by
scaling, translating, or rotating
the initial view. We first construct a $3 \times n$ matrix P, referred to
as the *coordinate matrix* of the view, whose columns are the coordinates of the n points of a view:

$$P = \begin{bmatrix} x_1 & x_2 & \cdots & x_n \\ y_1 & y_2 & \cdots & y_n \\ z_1 & z_2 & \cdots & z_n \end{bmatrix}.$$

For example, the coordinate matrix P corresponding to View 1 is the
3×12 matrix

$$\begin{bmatrix} 1.000 & .500 & -.500 & -1.000 & -.500 & .500 & .840 & .315 & -.210 & -.360 & -.210 & .315 \\ -.800 & -.800 & -.800 & -.800 & -.800 & -.800 & -.400 & .125 & .650 & .800 & .650 & .125 \\ .000 & -.866 & -.866 & .000 & .866 & .866 & .000 & -.546 & -.364 & .000 & .364 & .546 \end{bmatrix}$$

In the next few sections we show how to transform the coordinate
matrix P of a view to a new coordinate matrix P' corresponding to a
new view of the object. The straight line segments connecting the
various points move with the points as they are transformed. In
this way, each view is uniquely determined by its coordinate matrix
once we have specified which pairs of points in the original view
are to be connected by straight lines.

SCALING

The first type of transformation we consider consists of scal-
ing a view along the x, y, and z directions by factors of α, β, and
γ, respectively. By this we mean that if a point P_i has coordinates
(x_i, y_i, z_i) in the original view, it is to move to a new point P_i'
with coordinates $(\alpha x_i, \beta y_i, \gamma z_i)$ in the new view. This has the
effect of transforming a unit cube in the original view to a rectan-

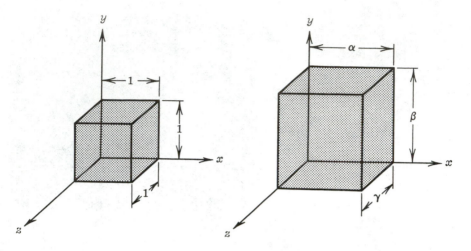

Figure 10.2

gular parallelepiped of dimensions $\alpha \times \beta \times \gamma$ (Fig. 10.2). Mathematically, this may be accomplished with matrix multiplication as follows: Define a 3×3 diagonal matrix

$$S = \begin{bmatrix} \alpha & 0 & 0 \\ 0 & \beta & 0 \\ 0 & 0 & \gamma \end{bmatrix}.$$

Then if a point P_i in the original view is represented by the column vector

$$\begin{bmatrix} x_i \\ y_i \\ z_i \end{bmatrix},$$

the transformed point P_i' is represented by the column vector

$$\begin{bmatrix} x_i' \\ y_i' \\ z_i' \end{bmatrix} = \begin{bmatrix} \alpha & 0 & 0 \\ 0 & \beta & 0 \\ 0 & 0 & \gamma \end{bmatrix} \begin{bmatrix} x_i \\ y_i \\ z_i \end{bmatrix}.$$

Using the coordinate matrix P which contains the coordinates of all n points of the original view as its columns, these n points can be transformed simultaneously to produce the coordinate matrix P' of the scaled view as follows:

$$SP = \begin{bmatrix} \alpha & 0 & 0 \\ 0 & \beta & 0 \\ 0 & 0 & \gamma \end{bmatrix} \begin{bmatrix} x_1 & x_2 & \cdots & x_n \\ y_1 & y_2 & \cdots & y_n \\ z_1 & z_2 & \cdots & z_n \end{bmatrix}$$

$$= \begin{bmatrix} \alpha x_1 & \alpha x_2 & \cdots & \alpha x_n \\ \beta y_1 & \beta y_2 & \cdots & \beta y_n \\ \gamma z_1 & \gamma z_2 & \cdots & \gamma z_n \end{bmatrix} = P' .$$

The new coordinate matrix can then be entered into the video display system to produce the new view of the object. As an example, View 2 is View 1 scaled by setting $\alpha = 1.8$, $\beta = 0.5$ and $\gamma = 3.0$. Notice that the scaling $\gamma = 3.0$ along the z-axis is not visible in View 2 since we only see the projection of the object onto the xy-plane.

View 2. View 1 scaled by $\alpha = 1.8$, $\beta = 0.5$, $\gamma = 3.0$.

TRANSLATION

We next consider the transformation of translating or displacing an object to a new position on the screen. Referring to Fig. 10.3, suppose we desire to change an existing view so that each point P_i with coordinates (x_i, y_i, z_i) moves to a new point P'_i with coordinates $(x_i + x_0, y_i + y_0, z_i + z_0)$. The vector

$$\begin{bmatrix} x_0 \\ y_0 \\ z_0 \end{bmatrix}$$

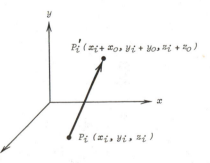

Figure 10.3

is called the *translation vector* of the transformation. By defining a $3 \times n$ matrix T as follows

$$T = \begin{bmatrix} x_0 & x_0 & \cdots & x_0 \\ y_0 & y_0 & \cdots & y_0 \\ z_0 & z_0 & \cdots & z_0 \end{bmatrix},$$

all n points of the view determined by the coordinate matrix P can be translated by matrix addition by means of the equation

$$P' = P + T.$$

The coordinate matrix P' then specifies the new coordinates of the n points. For example, if we wish to translate View 1 according to the translation vector

$$\begin{bmatrix} 1.2 \\ 0.4 \\ 1.7 \end{bmatrix},$$

View 3. View 1 translated by $x_0 = 1.2$, $y_0 = 0.4$, $z_0 = 1.7$.

the result is View 3. Notice, again, that the translation $z_0 = 1.7$ along the z-axis does not show up explicitly in View 3.

In Exercise 10.7, a technique of performing translations by matrix multiplication rather than by matrix addition as above is explained.

ROTATION

A more complicated type of transformation is a rotation of a view about one of the three coordinate axes. We begin with a rotation about the z-axis (the axis perpendicular to the screen) through an angle θ. Given a point P_i in the original view with coordinates (x_i, y_i, z_i), we wish to compute the new coordinates (x_i', y_i', z_i') of the rotated point P_i'. Referring to Fig. 10.4 and using a little trigonometry, the reader should be able to derive the following:

$$x_i' = \rho\cos(\phi+\theta) = \rho\cos\phi\cos\theta - \rho\sin\phi\sin\theta = x_i\cos\theta - y_i\sin\theta$$
$$y_i' = \rho\sin(\phi+\theta) = \rho\cos\phi\sin\theta + \rho\sin\phi\cos\theta = x_i\sin\theta + y_i\cos\theta$$
$$z_i' = z_i.$$

These equations may be written in
matrix form as

$$\begin{bmatrix} x_i' \\ y_i' \\ z_i' \end{bmatrix} = \begin{bmatrix} \cos\theta & -\sin\theta & 0 \\ \sin\theta & \cos\theta & 0 \\ 0 & 0 & 1 \end{bmatrix} \begin{bmatrix} x_i \\ y_i \\ z_i \end{bmatrix}.$$

If we let R denote the 3×3 matrix
in this equation, all n points may
be rotated by the matrix product

$$P' = RP$$

to yield the coordinate matrix P'
of the rotated view.

Rotations about the x- and
y-axes can be accomplished analogously, and the resulting rotation
matrices are given on the following page. Three new views of the
truncated pyramid are also displayed there, corresponding to
rotations of View 1 about the x-,
y-, and z-axes, each through an
angle of 90°.

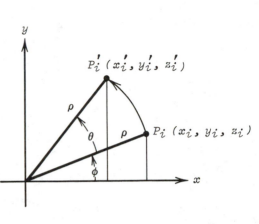

Figure 10.4

Rotations about the three coordinate axes may be combined to
give oblique views of an object. For example, View 7 is View 1
rotated first about the x-axis through 30°, then about y-axis
through -70°, and finally about the z-axis through -27°. Mathematically, these three successive rotations can be embodied in the
single transformation equation $P' = RP$ where R is the product of
three individual rotation matrices as follows:

$$\begin{bmatrix} \cos(-27°) & -\sin(-27°) & 0 \\ \sin(-27°) & \cos(-27°) & 0 \\ 0 & 0 & 1 \end{bmatrix} \begin{bmatrix} \cos(-70°) & 0 & \sin(-70°) \\ 0 & 1 & 0 \\ -\sin(-70°) & 0 & \cos(-70°) \end{bmatrix} \begin{bmatrix} 1 & 0 & 0 \\ 0 & \cos(30°) & -\sin(30°) \\ 0 & \sin(30°) & \cos(30°) \end{bmatrix}$$

$$= \begin{bmatrix} .305 & -.025 & -.952 \\ -.155 & .985 & -.076 \\ .940 & .171 & .296 \end{bmatrix}$$

ROTATION ABOUT THE X-AXIS

$$\begin{bmatrix} 1 & 0 & 0 \\ 0 & \cos\theta & -\sin\theta \\ 0 & \sin\theta & \cos\theta \end{bmatrix}$$

ROTATION ABOUT THE Y-AXIS

$$\begin{bmatrix} \cos\theta & 0 & \sin\theta \\ 0 & 1 & 0 \\ -\sin\theta & 0 & \cos\theta \end{bmatrix}$$

ROTATION ABOUT THE Z-AXIS

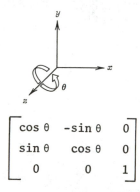

$$\begin{bmatrix} \cos\theta & -\sin\theta & 0 \\ \sin\theta & \cos\theta & 0 \\ 0 & 0 & 1 \end{bmatrix}$$

View 4. View 1 rotated 90° about x-axis.

View 5. View 1 rotated 90° about y-axis.

View 6. View 1 rotated 90° about z-axis.

As a final illustration, in View 8 we have two separate views of the truncated pyramid which constitute a stereoscopic pair. They were produced by first rotating View 7 about the y-axis through an angle of -3° and translating it to the right, then rotating the same View 7 about the y-axis through an angle of +3° and translating it to the left. The translation distances were chosen so that the stereoscopic views are about 2½" apart — the approximate distance between a pair of eyes.

View 7. Oblique view of truncated pyramid.

View 8. Stereoscopic figure of truncated pyramid. The three-dimensionality of the diagram can be seen by holding the book about one foot away and focusing on a distant object. Then by shifting gaze to View 8 without refocusing, the two views of the stereoscopic pair can be made to merge together and produce the desired effect.

EXERCISES

10.1 View 9 is a view of a square with vertices $(0,0,0)$, $(1,0,0)$, $(1,1,0)$, and $(0,1,0)$.

(a) What is the coordinate matrix of View 9?

(b) What is the coordinate matrix of View 9 after it is scaled by a factor $1\frac{1}{2}$ in the x-direction and $\frac{1}{2}$ in the y-direction? Draw a sketch of the scaled view.

(c) What is the coordinate matrix of View 9 after it is translated by the vector

Draw a sketch of the translated view.

(d) What is the coordinate matrix of View 9 after it is rotated through an angle of -30° about the z-axis. Draw a sketch of the rotated view.

10.2 (a) If the coordinate matrix of View 9 is multiplied by the matrix

$$\begin{bmatrix} 1 & \frac{1}{2} & 0 \\ 0 & 1 & 0 \\ 0 & 0 & 1 \end{bmatrix}$$

the result is the coordinate matrix of View 10. Such a transformation is called a *shear along the x-axis of factor $\frac{1}{2}$ with respect to the y-coordinate*. Show that under such a transformation a point with coordinates (x_i, y_i, z_i) has new coordinates $(x_i + \frac{1}{2}y_i, y_i, z_i)$.

(b) What are the coordinates of the four vertices of the sheared square in View 10?

(c) The matrix

$$\begin{bmatrix} 1 & 0 & 0 \\ .6 & 1 & 0 \\ 0 & 0 & 1 \end{bmatrix}$$

View 9. Square with vertices (0,0,0), (1,0,0), (1,1,0), and (0,1,0). (Ex. 10.1 and 10.2) ▶

View 10. View 9 sheared along x-axis by ½ with respect to the y-coordinate. (Ex. 10.2) ◀

View 11. View 1 sheared along y-axis by .6 with respect to the x-coordinate. (Ex. 10.2) ▶

View 12. View 1 reflected about the xz-plane. (Ex. 10.3)

View 13. View 1 scaled, translated, and rotated. (Ex. 10.4)

View 14. View 1 scaled, translated, and rotated. (Ex. 10.5)

determines a *shear along the y-axis of factor .6 with respect to the x-coordinate* (e.g. View 11). Sketch a view of the square in View 9 after such a shearing transformation and find the new coordinates of its four vertices.

10.3 (a) The *reflection about the xz-plane* is defined as the transformation which takes a point (x_i, y_i, z_i) to the point $(x_i, -y_i, z_i)$ (e.g. View 12). If P and P' and the coordinate matrices of a view and its reflection about the xz-plane, respectively, find a matrix M such that $P' = MP$.

(b) Analogous to part (a), define the *reflection about the yz-plane* and construct the corresponding transformation matrix. Draw a sketch of View 1 reflected about the yz-plane.

(c) Analogous to part (a), define the *reflection about the xy-plane* and construct the corresponding transformation matrix. Draw a sketch of View 1 reflected about the xy-plane.

10.4 (a) View 13 is View 1 subject to the following five transformations:

1. Scale by a factor of 1/2 in the x-direction, 2 in the y-direction, and 1/3 in the z-direction.
2. Translate 1/2 unit in the x-direction.
3. Rotate 20° about the x-axis.
4. Rotate -45° about the y-axis.
5. Rotate 90° about the z-axis.

Construct the five matrices M_1, M_2, M_3, M_4, and M_5 associated with these five transformations.

(b) If P is the coordinate matrix of View 1 and P' is the coordinate matrix of View 13, express P' in terms of M_1, M_2, M_3, M_4, M_5, and P.

10.5 (a) View 14 is View 1 subject to the following seven transformations:

1. Scale by a factor of .3 in the x-direction and .5 in the y-direction.
2. Rotate 45° about the x-axis.
3. Translate 1 unit in the x-direction.
4. Rotate 35° about the y-axis.
5. Rotate -45° about the z-axis.
6. Translate 1 unit in the z-direction.
7. Scale by a factor of 2 in the x-direction.

Construct the seven matrices $M_1, M_2, \ldots, M_7$ associated with these seven transformations.

(b) If P is the coordinate matrix of View 1 and P' is the co-ordinate matrix of View 14, express P' in terms of M_1, $M_2, \ldots, M_7$, and P.

10.6 Suppose a view with coordinate matrix P is to be rotated through an angle θ about an axis through the origin and specified by two angles α and β (Fig. 10.5). If P' is the coordinate matrix of the rotated view, find rota-tion matrices R_1, R_2, R_3, R_4, and R_5 such that

$$P' = R_5 R_4 R_3 R_2 R_1 P.$$

Hint: The desired rotation can be accomplished in the following five steps:

Figure 10.5

1. Rotate through an angle of β about the y-axis
2.   "   "   "   "   " α "   " z-axis.
3.   "   "   "   "   " θ "   " y-axis.
4.   "   "   "   "   " $-\alpha$ "   " z-axis.
5.   "   "   "   "   " $-\beta$ "   " y-axis.

10.7 This exercise illustrates a technique of translating a point with coordinates (x_i, y_i, z_i) to a point with coordinates $(x_i + x_0, y_i + y_0, z_i + z_0)$ by matrix multiplication rather than by matrix addition.

(a) Let the point (x_i, y_i, z_i) be associated with the column vector

$$v_i = \begin{bmatrix} x_i \\ y_i \\ z_i \\ 1 \end{bmatrix}$$

and let the point $(x_i + x_0, y_i + y_0, z_i + z_0)$ be associated with the column vector

$$v'_i = \begin{bmatrix} x_i + x_0 \\ y_i + y_0 \\ z_i + z_0 \\ 1 \end{bmatrix}.$$

Find a 4×4 matrix M such that $v'_i = Mv_i$.

(b) Find the specific 4×4 matrix of the above form that will effect the translation of the point $(4, -2, 3)$ to the point $(-1, 7, 0)$.

10.8 For the three rotation matrices given on page 128, show that

$$R^{-1} = R^t.$$

(A matrix with this property is called an *orthogonal matrix*.)

Equilibrium Temperature Distributions

*The equilibrium temperature distribution
within a trapezoidal plate is found when the
temperature around the edges of the plate is
specified. The problem is reduced to solving
a linear system of equations and an iterative
technique for solving the system is described.
A "random walk" approach to the problem is
also discussed.*

PREREQUISITES: Linear systems
Matrices
Intuitive understanding of limits

INTRODUCTION

Suppose we are given a thin trapezoidal plate (Fig. 11.1(a)),
whose two faces are insulated from heat. Suppose we are also given
the temperature along the four edges of the plate. For example, let
the temperature be constant on each edge with values of 0°, 0°, 1°,
and 2°, as in the diagram. After a period of time, the temperature
inside the plate will stabilize. Our objective in this chapter is

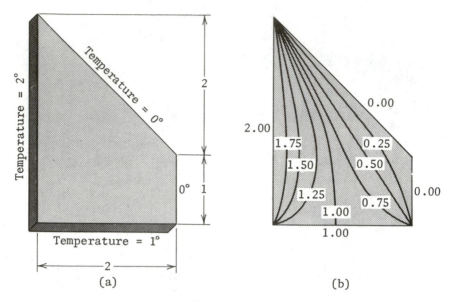

Figure 11.1

to determine this equilibrium temperature distribution at the points inside the plate. As we shall see, the interior equilibrium temperature is completely determined by the *boundary data*, that is, the temperature along the edges of the plate.

The equilibrium temperature distribution can be visualized by the use of curves which connect points of equal temperature. Such curves are called *isotherms* of the temperature distribution. In Fig. 11.1(b) we have sketched a few isotherms using information we derive later in the chapter.

Though all of our calculations will be for the trapezoidal plate illustrated, our techniques generalize easily to a plate of any shape. They also generalize to the problem of finding the temperature within a three-dimensional body. In fact, our "plate" could be the cross-section of some solid object if the flow of heat perpendicular to the cross-section is negligible. For example, Fig. 11.1 could represent the cross-section of a long dam. The dam is exposed to three different temperatures: the temperature of the ground at its base, the temperature of the water on one side, and the temperature of the air on the other side. A knowledge of the temperature distribution inside the dam is necessary to determine the thermal stresses which it is subjected to.

In the next section, we begin with a certain thermodynamic principle which characterizes the temperature distribution we are seeking.

THE MEAN-VALUE PROPERTY

There are many different ways to obtain a mathematical model for our problem. The approach we shall use is based on the following property of equilibrium temperature distributions:

The Mean-Value Property

Let a plate be in thermal equilibrium and let P be a point inside the plate. Then if C is any circle with center at P which is completely contained in the plate, the temperature at P is the average value of the temperature on the circle (Fig. 11.2).

Figure 11.2

This property is a consequence of certain basic laws of molecular motion and we shall not attempt to derive it. Basically, this property states that in equilibrium thermal energy tends to distribute itself as evenly as possible consistent with the boundary conditions. It can be shown that the mean-value property uniquely determines the equilibrium temperature distribution of a plate.

Unfortunately, the determination of the equilibrium temperature distribution from the mean-value property is not an easy matter. However, if we restrict ourselves to finding the temperature only at a finite set of points within the plate, the problem can be reduced to solving a linear system. We shall pursue this idea in the next section.

DISCRETE FORMULATION OF THE PROBLEM

Let us overlay our trapezoidal plate with a succession of finer and finer square nets or meshes (Fig. 11.3). In (a) we have a rather coarse net; in (b) we have a net with half the spacing as in (a); and in (c) we have a net with the spacing again reduced by half. The points of intersection of the net lines are called *mesh points*. We classify them as *boundary mesh points* if they fall on the boundary of the plate or *interior mesh points* if they lie in the interior of the plate. For the three net spacings we have chosen, there are 1, 9, and 49 interior mesh points respectively.

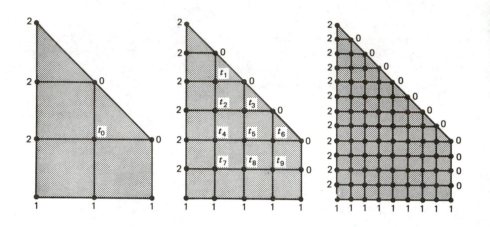

1 Interior Meshpoint   9 Interior Meshpoints   49 Interior Meshpoints

(a)   (b)   (c)

Figure 11.3

In the discrete formulation of our problem, we try to find the temperature only at the interior mesh points of some particular net. For a rather fine net, such as in (c), this will provide an excellent picture of the temperature distribution throughout the entire plate.

At the boundary mesh points, the temperature is given by the boundary data. (In Fig. 11.3 we have labeled all of the boundary mesh points with their corresponding temperatures.) At the interior mesh points, we shall apply the following discrete version of the mean - value property:

Discrete Mean - Value Property

At each interior mesh point, the temperature is the average of the temperatures at the four neighboring mesh points.

This discrete version is a reasonable approximation to the true mean - value property. But since it is only an approximation, it will provide only an approximation to the true temperatures at the interior mesh points. However, the approximations will get better as the mesh spacing decreases. In fact, as the mesh spacing approaches zero, the approximations approach the exact temperature distribution, a fact proved in advanced courses in Numerical

Analysis. We shall illustrate this convergence by computing the approximate temperatures at the mesh points for the three mesh spacings given in Fig. 11.3.

Case (a) of Fig. 11.3 is quite simple since there is only one interior mesh point. If we let t_0 be the temperature at this mesh point, the discrete mean-value property immediately gives

$$t_0 = \tfrac{1}{4}(2 + 1 + 0 + 0) = .75.$$

In case (b), let us label the temperatures at the nine interior mesh points $t_1, t_2, \ldots, t_9$, as in Fig. 11.3(b). (The particular ordering is not important.) By applying the discrete mean-value property successively to each of these nine mesh points, we obtain the following nine equations:

$$
\begin{aligned}
t_1 &= \tfrac{1}{4}(t_2 + 2 + 0 + 0) \\
t_2 &= \tfrac{1}{4}(t_1 + t_3 + t_4 + 2) \\
t_3 &= \tfrac{1}{4}(t_2 + t_5 + 0 + 0) \\
t_4 &= \tfrac{1}{4}(t_2 + t_5 + t_7 + 2) \\
t_5 &= \tfrac{1}{4}(t_3 + t_4 + t_6 + t_8) \\
t_6 &= \tfrac{1}{4}(t_5 + t_9 + 0 + 0) \\
t_7 &= \tfrac{1}{4}(t_4 + t_8 + 1 + 2) \\
t_8 &= \tfrac{1}{4}(t_5 + t_7 + t_9 + 1) \\
t_9 &= \tfrac{1}{4}(t_6 + t_8 + 1 + 0)
\end{aligned}
\tag{11.1}
$$

This is a linear system of nine equations in nine unknowns. We may rewrite it in matrix form as

$$t = Mt + b \tag{11.2}$$

where

$$
t = \begin{bmatrix} t_1 \\ t_2 \\ t_3 \\ t_4 \\ t_5 \\ t_6 \\ t_7 \\ t_8 \\ t_9 \end{bmatrix}, \quad
M = \begin{bmatrix}
0 & \tfrac{1}{4} & 0 & 0 & 0 & 0 & 0 & 0 & 0 \\
\tfrac{1}{4} & 0 & \tfrac{1}{4} & \tfrac{1}{4} & 0 & 0 & 0 & 0 & 0 \\
0 & \tfrac{1}{4} & 0 & 0 & \tfrac{1}{4} & 0 & 0 & 0 & 0 \\
0 & \tfrac{1}{4} & 0 & 0 & \tfrac{1}{4} & 0 & \tfrac{1}{4} & 0 & 0 \\
0 & 0 & \tfrac{1}{4} & \tfrac{1}{4} & 0 & \tfrac{1}{4} & 0 & \tfrac{1}{4} & 0 \\
0 & 0 & 0 & 0 & \tfrac{1}{4} & 0 & 0 & 0 & \tfrac{1}{4} \\
0 & 0 & 0 & \tfrac{1}{4} & 0 & 0 & 0 & \tfrac{1}{4} & 0 \\
0 & 0 & 0 & 0 & \tfrac{1}{4} & 0 & \tfrac{1}{4} & 0 & \tfrac{1}{4} \\
0 & 0 & 0 & 0 & 0 & \tfrac{1}{4} & 0 & \tfrac{1}{4} & 0
\end{bmatrix}, \quad
b = \begin{bmatrix} 1/2 \\ 1/2 \\ 0 \\ 1/2 \\ 0 \\ 0 \\ 3/4 \\ 1/4 \\ 1/4 \end{bmatrix}.
$$

To solve Eq. (11.2), let us write it as

$$(I - M)\mathbf{t} = \mathbf{b}.$$

The solution for **t** is thus

$$\mathbf{t} = (I - M)^{-1}\mathbf{b} \qquad (11.3)$$

as long as the matrix $(I - M)$ is invertible. This is indeed the case, and the solution for **t** as calculated by (11.3) is

$$\mathbf{t} = \begin{bmatrix} 0.7846 \\ 1.1383 \\ 0.4719 \\ 1.2967 \\ 0.7491 \\ 0.3265 \\ 1.2995 \\ 0.9014 \\ 0.5570 \end{bmatrix}. \qquad (11.4)$$

Figure 11.4

Figure 11.4 is a diagram of the plate with the nine interior mesh points labeled with their temperatures as given by this solution.

For case (c) of Fig. 11.3, we repeat this same procedure. We label the temperatures at the 49 interior mesh points as $t_1, t_2, \ldots t_{49}$ in some manner. For example, we may begin at the top of the plate and proceed from left to right along each row of mesh points. Applying the discrete mean-value property to each mesh point gives a linear system of 49 equations in 49 unknowns:

$$t_1 = \tfrac{1}{4}(t_2 + 2 + 0 + 0)$$
$$t_2 = \tfrac{1}{4}(t_1 + t_3 + t_4 + 2)$$
$$\vdots \qquad\qquad (11.5)$$
$$t_{48} = \tfrac{1}{4}(t_{41} + t_{47} + t_{49} + 1)$$
$$t_{49} = \tfrac{1}{4}(t_{42} + t_{48} + 0 + 1).$$

In matrix form, Eqs. (11.5) are

$$t = Mt + b$$

where **t** and **b** are column vectors with 49 entries and M is a 49×49 matrix. As in (11.3), the solution for **t** is

$$t = (I - M)^{-1}b. \tag{11.6}$$

In Fig. 11.5 we display the temperatures at the 49 mesh points found by Eq. (11.6). The nine unshaded temperatures in this figure fall on the same mesh points of Fig. 11.4. In Table 11.1 we compare the temperatures at these nine common mesh points for the three different mesh spacings used.

Temperatures at Common Mesh Points		
Case (a)	Case (b)	Case (c)
t_1 —	0.7846	0.8048
t_2 —	1.1383	1.1533
t_3 —	0.4719	0.4778
t_4 —	1.2967	1.3078
t_5 0.7500	0.7491	0.7513
t_6 —	0.3265	0.3157
t_7 —	1.2995	1.3042
t_8 —	0.9014	0.9032
t_9 —	0.5570	0.5554

Table 11.1

Knowing that the temperatures of the discrete problem approach the exact temperatures as the mesh spacing decreases, we may surmise that the 9 temperatures obtained in case (c) are closer to the exact values than those in case (b).

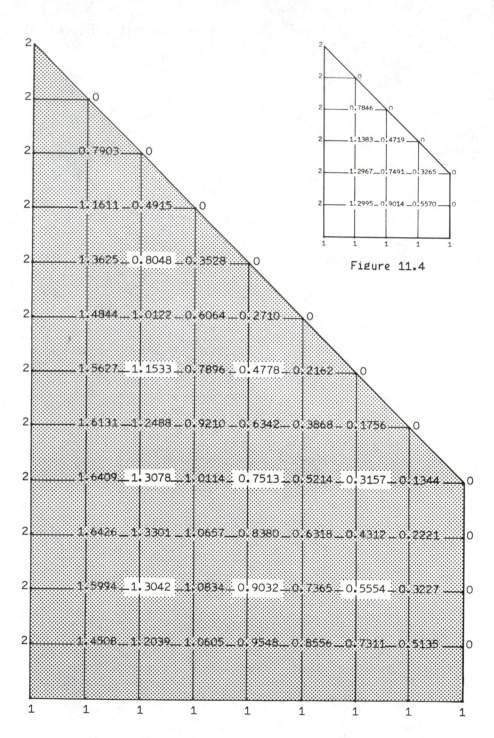

Figure 11.4

Figure 11.5

A NUMERICAL TECHNIQUE

To obtain the 49 temperatures in case (c) above, it was necessary to solve a linear system with 49 unknowns. A finer net might involve a linear system with hundreds or even thousands of unknowns. Exact algorithms for the solutions of such large systems are impractical, and for this reason we shall now discuss a numerical technique for the practical solution of these systems.

To describe this technique, let us look again at Eq. (11.2):

$$t = Mt + b. \tag{11.7}$$

The vector t we are seeking appears on both sides of this equation. We consider a way of generating better and better approximations to the vector solution t. For the initial approximation $t^{(0)}$ we may take $t^{(0)} = 0$ if no better choice is available. If we substitute $t^{(0)}$ into the righthand side of (11.7) and label the resulting lefthand side as $t^{(1)}$, we have

$$t^{(1)} = Mt^{(0)} + b. \tag{11.8}$$

Usually $t^{(1)}$ is a better approximation to the solution than is $t^{(0)}$. If we substitute $t^{(1)}$ into the righthand side of (11.7) we generate another approximation, which we label as $t^{(2)}$:

$$t^{(2)} = Mt^{(1)} + b. \tag{11.9}$$

Continuing in this way, we generate a sequence of approximations as follows:

$$\begin{aligned}
t^{(1)} &= Mt^{(0)} + b \\
t^{(2)} &= Mt^{(1)} + b \\
t^{(3)} &= Mt^{(2)} + b \\
&\ \vdots \\
t^{(n)} &= Mt^{(n-1)} + b \\
&\ \vdots
\end{aligned} \tag{11.10}$$

Hopefully, this sequence of approximations $t^{(0)}$, $t^{(1)}$, $t^{(2)}$, ... converges to the exact solution of (11.7). We do not have the space in this book to go into the theoretical considerations necessary to show this. However, suffice it to say that for the particular problem we are considering, the sequence converges to the exact solution for any mesh size and for any initial approximation $t^{(0)}$.

This technique of generating successive approximations to the solution of a linear system is called *Jacobi iteration*, and the approximations themselves are called *iterates*. As a numerical example, let us apply Jacobi iteration to the calculation of the nine mesh point temperatures of case (b). Setting $t^{(0)} = 0$, we have from Eq. (11.2):

$$t^{(1)} = Mt^{(0)} + b = M0 + b = b = \begin{bmatrix} .5000 \\ .5000 \\ .0000 \\ .5000 \\ .0000 \\ .0000 \\ .7500 \\ .2500 \\ .2500 \end{bmatrix},$$

$$t^{(2)} = Mt^{(1)} + b$$

$$= \begin{bmatrix} 0 & \frac{1}{4} & 0 & 0 & 0 & 0 & 0 & 0 & 0 \\ \frac{1}{4} & 0 & \frac{1}{4} & \frac{1}{4} & 0 & 0 & 0 & 0 & 0 \\ 0 & \frac{1}{4} & 0 & 0 & \frac{1}{4} & 0 & 0 & 0 & 0 \\ 0 & \frac{1}{4} & 0 & 0 & \frac{1}{4} & 0 & \frac{1}{4} & 0 & 0 \\ 0 & 0 & \frac{1}{4} & \frac{1}{4} & 0 & \frac{1}{4} & 0 & \frac{1}{4} & 0 \\ 0 & 0 & 0 & 0 & \frac{1}{4} & 0 & 0 & 0 & \frac{1}{4} \\ 0 & 0 & 0 & \frac{1}{4} & 0 & 0 & 0 & \frac{1}{4} & 0 \\ 0 & 0 & 0 & 0 & \frac{1}{4} & 0 & \frac{1}{4} & 0 & \frac{1}{4} \\ 0 & 0 & 0 & 0 & 0 & \frac{1}{4} & 0 & \frac{1}{4} & 0 \end{bmatrix} \begin{bmatrix} .5000 \\ .5000 \\ .0000 \\ .5000 \\ .0000 \\ .0000 \\ .7500 \\ .2500 \\ .2500 \end{bmatrix} + \begin{bmatrix} .5000 \\ .5000 \\ .0000 \\ .5000 \\ .0000 \\ .0000 \\ .7500 \\ .2500 \\ .2500 \end{bmatrix} = \begin{bmatrix} .6250 \\ .7500 \\ .1250 \\ .8125 \\ .1875 \\ .0625 \\ .9375 \\ .5000 \\ .3125 \end{bmatrix}.$$

Some additional iterates are

$$
t^{(3)} = \begin{bmatrix} 0.6875 \\ 0.8906 \\ 0.2344 \\ 0.9688 \\ 0.3750 \\ 0.1250 \\ 1.0781 \\ 0.6094 \\ 0.3906 \end{bmatrix}, \quad
t^{(10)} = \begin{bmatrix} 0.7791 \\ 1.1230 \\ 0.4573 \\ 1.2770 \\ 0.7236 \\ 0.3131 \\ 1.2848 \\ 0.8827 \\ 0.5446 \end{bmatrix}, \quad
t^{(20)} = \begin{bmatrix} 0.7845 \\ 1.1380 \\ 0.4716 \\ 1.2963 \\ 0.7486 \\ 0.3263 \\ 1.2992 \\ 0.9010 \\ 0.5567 \end{bmatrix}, \quad
t^{(30)} = \begin{bmatrix} 0.7846 \\ 1.1383 \\ 0.4719 \\ 1.2967 \\ 0.7491 \\ 0.3265 \\ 1.2995 \\ 0.9014 \\ 0.5570 \end{bmatrix}.
$$

All iterates beginning with the thirtieth are equal to $t^{(30)}$ to four decimal places. Consequently, $t^{(30)}$ is the exact solution to four decimal places. This agrees with our previous result given in Eq. (11.4).

The Jacobi iteration scheme applied to the linear system (11.5) with 49 unknowns produced iterates which begin repeating to four decimal places after 119 iterations. Thus, $t^{(119)}$ would provide the 49 temperatures of case (c) correct to four decimal places.

A MONTE CARLO TECHNIQUE

In this section, we describe a so-called *Monte Carlo technique* for computing the temperature at a single interior mesh point of the discrete problem without having to compute the temperatures at the remaining interior mesh points. First we define a *discrete random walk along the net*. By this we mean a directed path along the net lines (Fig. 11.6) which joins a succession of mesh points such that the direction of departure from each mesh point is chosen at random. Each of the four possible directions of departure from each mesh point along the path is to be equally probable.

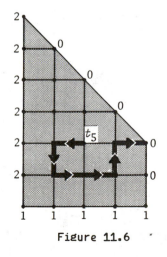

Figure 11.6

By the use of random walks, we can compute the temperature at a specified interior mesh point on the basis of the following property:

Random Walk Property

Let W_1, W_2, ..., W_n be a succession of random walks, all of which begin at a specified interior mesh point. Let t_1^*, t_2^*, ..., t_n^* be the temperatures at the boundary mesh points first encountered along each of these random walks. Then the average value $(t_1^* + t_2^* + \cdots + t_n^*)/n$ of these boundary temperatures approaches the temperature at the specified interior mesh point as the number of random walks n increases without bound.

This property is a consequence of the discrete mean-value property which the mesh point temperatures satisfy. The proof of the random walk property involves elementary concepts from probability theory and we shall not give it here.

In Table 11.2 we display the results of a large number of computer-generated random walks for the evaluation of the temperature t_5 of the nine-point mesh of case (b) (Fig. 11.6). The first column lists the number n of the random walk. The second column lists the temperature t_n^* of the boundary point first encountered along the corresponding random walk. The last column contains the cumulative average of the boundary temperatures encountered along the n random walks.

n	t_n^*	$(t_1^* + \cdots + t_n^*)/n$	n	t_n^*	$(t_1^* + \cdots + t_n^*)/n$
1	1	1.0000	20	1	0.9500
2	2	1.5000	30	0	0.8000
3	1	1.3333	40	0	0.8250
4	0	1.0000	50	2	0.8400
5	2	1.2000	100	0	0.8300
6	0	1.0000	150	1	0.8000
7	2	1.1429	200	0	0.8050
8	0	1.0000	250	1	0.8240
9	2	1.1111	500	1	0.7860
10	0	1.0000	1000	0	0.7550

Table 11.2

Thus, after 1000 random walks we have the approximation $t_5 \simeq .7550$. This compares with the exact value $t_5 = .7491$ we had previously evaluated. As can be seen, the convergence to the exact value is not too rapid.

EXERCISES

11.1 A plate in the form of a circular disk has boundary temperatures of 0° on half of its circumference and 1° on the remaining half of its circumference. A net with four interior mesh points is overlaid on the disk (Fig. 11.7).

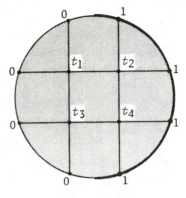

Figure 11.7

(a) Using the discrete mean-value property, write the 4×4 linear system $t = Mt + b$ which determines the approximate temperatures at the four interior mesh points.

(b) Solve the linear system in part (a).

(c) Use the Jacobi interation scheme with $t^{(0)} = 0$ to generate the iterates $t^{(1)}$, $t^{(2)}$, $t^{(3)}$, $t^{(4)}$, and $t^{(5)}$ for the linear system in part (a). What is the "error vector" $t^{(5)} - t$ where t is the solution found in part (b)?

(d) By certain advanced methods, it can be determined that the exact temperatures at the four mesh points are $t_1 = t_3 = .2371$ and $t_2 = t_4 = .7629$. What are the percentage errors in the values found in part (b)?

11.2 Using the exact mean-value property (page 137) find the exact equilibrium temperature at the center of the disk in Exercise 11.1.

11.3 Calculate the first two iterates $t^{(1)}$ and $t^{(2)}$ for case (b) with nine interior mesh points (Eq. (11.2)) when the initial iterate is chosen as

$$t^{(0)} = [1 \quad 1 \quad 1 \quad 1 \quad 1 \quad 1 \quad 1 \quad 1 \quad 1]^t.$$

11.4 The random walk illustrated in Fig. 11.6 may be described by six arrows

$$\leftarrow \ \downarrow \ \rightarrow \ \rightarrow \ \uparrow \ \rightarrow$$

which specify the directions of departure from the successive mesh points along the path. Below is an array of 100 computer-generated, randomly-oriented arrows arranged in a 10×10 array. Use these arrows to determine ten random walks to approximate the temperature t_5 as in Table 11.2. Proceed as follows:

(1) Take the last two digits of your telephone number. Use the last digit to specify a row and the other to specify a column.

(2) Go to the arrow in the array with that row and column number.

(3) Using this arrow as a starting point, move through the array of arrows as you would read a book (left to right and top to bottom). Beginning at the point labeled t_5 in Fig. 11.6 and using this sequence of arrows to specify a sequence of directions, move from mesh point to mesh point until you reach a boundary mesh point. This completes your first random walk. Record the temperature at the boundary mesh point. (If you reach the end of the arrow array, continue with the arrow in the upper lefthand corner.)

(4) Return to the interior mesh point labeled t_5 and begin where you left off in the arrow array and generate your next random walk. Repeat this process until you have completed ten random walks and have recorded ten boundary temperatures.

(5) Calculate the average of the ten boundary temperatures recorded. (The exact value is $t_5 = .7491$.)

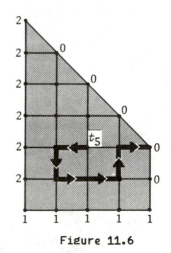

Figure 11.6

Some Applications in Genetics

Using the diagonalization of a matrix to compute its powers, the propagation of an inherited trait in successive generations is investigated.

PREREQUISITES: Eigenvalues and eigenvectors
Diagonalization of a matrix
Intuitive understanding of limits

INTRODUCTION

In this chapter, we shall examine the inheritance of traits in animals or plants. The inherited trait under consideration is assumed to be governed by a set of two genes, which we designate by the letters A and a. Under *autosomal inheritance* each individual in the population, of either sex, possesses two of these genes: the possible pairings being designated AA, Aa, and aa. This pair of genes is called the individual's *genotype*, and it determines how the trait controlled by the genes is manifested in the individual. For example, in snapdragons a set of two genes determines the color of the flower. Genotype AA produces red flowers, genotype Aa produces pink flowers, and genotype aa produces white flowers. In humans, eye coloration is controlled through autosomal inheritance. Genotypes AA and Aa have brown eyes, and genotype aa has blue eyes. In this case, we say that gene *A* *dominates* gene *a*, or that gene *a* is *recessive* to gene *A*, since genotype *Aa* has the same outward trait as genotype *AA*.

In addition to autosomal inheritance, we shall also discuss *X-linked inheritance*. In this type of inheritance the male of the species possesses only one of the two possible genes (*A* or *a*), and the female possesses a pair of the two genes (*AA*, *Aa*, or *aa*). In humans, color blindness, hereditary baldness, hemophilia, and muscular dystrophy, to name a few, are traits controlled by X-linked inheritance.

In the next sections, we explain the manner in which the genes of the parents are passed on to their offspring for the two types of inheritance. We construct matrix models which give the probable genotypes of the offspring in terms of the genotypes of the parents, and use these matrix models to follow the genotype distribution of a population through successive generations.

Autosomal Inheritance

In autosomal inheritance, an individual inherits one gene from each of its parents' pairs of genes to form its own particular pair. As far as we know, it is a matter of chance which of the two genes a parent passes on to the offspring. Thus, if one parent is of genotype *Aa*, it is equally likely that the offspring will inherit the *A* gene or the *a* gene from that parent. If one parent is of genotype *aa* and the other parent is of genotype *Aa*, the offspring will always receive an *a* gene from the *aa* parent, and will receive either an *A* gene or an *a* gene, with equal probability, from the *Aa* parent. Consequently, the offspring has equal probability of being genotype *AA* or *Aa*. In Table 12.1, we list the probabilities of the possible genotypes of the offspring for all possible combinations of the genotypes of the parents.

			Genotypes of Parents					
			AA - AA	*AA - Aa*	*AA - aa*	*Aa - Aa*	*Aa - aa*	*aa - aa*
Genotype of Offspring		*AA*	1	½	0	¼	0	0
		Aa	0	½	1	½	½	0
		aa	0	0	0	¼	½	1

Table 12.1

EXAMPLE 12.1 Suppose a farmer has a large population of plants consisting of some distribution of all three possible genotypes AA, Aa, and aa. He desires to undertake a breeding program in which each plant in the population is always fertilized with a plant of genotype AA. We want to derive an expression for the distribution of the three possible genotypes in the population after any number of generations.

For $n = 0, 1, 2, \ldots$, let us set

a_n = fraction of plants of genotype AA in n-th generation,

b_n = fraction of plants of genotype Aa in n-th generation,

c_n = fraction of plants of genotype aa in n-th generation.

Thus, a_0, b_0, and c_0 specify the initial distribution of the genotypes. We also have that

$$a_n + b_n + c_n = 1 \quad \text{for} \quad n = 0, 1, 2, \ldots$$

From Table 12.1, we may determine the genotype distribution of each generation from the genotype distribution of the preceeding generation by the following equations:

$$a_n = a_{n-1} + \tfrac{1}{2} b_{n-1}$$

$$b_n = c_{n-1} + \tfrac{1}{2} b_{n-1} \qquad n = 1, 2, \ldots \quad (12.1)$$

$$c_n = 0.$$

For example, the first of these three equations states that all the offspring of a plant of genotype AA will be of genotype AA under this breeding program, and half of the offspring of a plant of genotype Aa will be of genotype AA.

Equations (12.1) can be written in matrix notation as

$$x^{(n)} = M x^{(n-1)} \qquad n = 1, 2, \ldots \quad (12.2)$$

where

$$x^{(n)} = \begin{bmatrix} a_n \\ b_n \\ c_n \end{bmatrix}, \quad x^{(n-1)} = \begin{bmatrix} a_{n-1} \\ b_{n-1} \\ c_{n-1} \end{bmatrix}, \quad \text{and} \quad M = \begin{bmatrix} 1 & \tfrac{1}{2} & 0 \\ 0 & \tfrac{1}{2} & 1 \\ 0 & 0 & 0 \end{bmatrix}.$$

Notice that the three columns of the matrix M are the same as the first three columns of Table 12.1.

From Eq. (12.2) it follows that

$$x^{(n)} = Mx^{(n-1)} = M^2 x^{(n-2)} = \cdots = M^n x^{(0)}. \qquad (12.3)$$

Consequently, if we can find an explicit expression for M^n, we can use Eq. (12.3) to obtain an explicit expression $x^{(n)}$. To find an explicit expression for M^n, we first diagonalize M. That is, we find an invertible matrix P and a diagonal matrix D such that

$$M = PDP^{-1}. \qquad (12.4)$$

With such a diagonalization, we then have (see Exercise 12.1)

$$M^n = PD^n P^{-1} \qquad \text{for } n = 1, 2, \ldots$$

where

$$D^n = \begin{bmatrix} d_1 & 0 & 0 & \cdots & 0 \\ 0 & d_2 & 0 & \cdots & 0 \\ \vdots & \vdots & \vdots & & \vdots \\ 0 & 0 & 0 & & d_k \end{bmatrix}^n = \begin{bmatrix} d_1^n & 0 & 0 & \cdots & 0 \\ 0 & d_2^n & 0 & \cdots & 0 \\ \vdots & \vdots & \vdots & & \vdots \\ 0 & 0 & 0 & \cdots & d_k^n \end{bmatrix}.$$

The diagonalization of M is accomplished by finding its eigenvalues and corresponding eigenvectors. These are easily found to be (verify):

Eigenvalues: $\qquad \lambda_1 = 1, \qquad \lambda_2 = \tfrac{1}{2}, \qquad \lambda_3 = 0$

Corresponding
Eigenvectors: $\quad e_1 = \begin{bmatrix} 1 \\ 0 \\ 0 \end{bmatrix}, \qquad e_2 = \begin{bmatrix} 1 \\ -1 \\ 0 \end{bmatrix}, \qquad e_3 = \begin{bmatrix} 1 \\ -2 \\ 1 \end{bmatrix}.$

In Eq. (12.4), we then have

$$D = \begin{bmatrix} \lambda_1 & 0 & 0 \\ 0 & \lambda_2 & 0 \\ 0 & 0 & \lambda_3 \end{bmatrix} = \begin{bmatrix} 1 & 0 & 0 \\ 0 & \tfrac{1}{2} & 0 \\ 0 & 0 & 0 \end{bmatrix}$$

and

$$P = \left[\mathbf{e}_1 \mid \mathbf{e}_2 \mid \mathbf{e}_3 \right] = \begin{bmatrix} 1 & 1 & 1 \\ 0 & -1 & -2 \\ 0 & 0 & 1 \end{bmatrix}.$$

Thus, we have

$$\mathbf{x}^{(n)} = PD^n P^{-1} \mathbf{x}^{(0)} = \begin{bmatrix} 1 & 1 & 1 \\ 0 & -1 & -2 \\ 0 & 0 & 1 \end{bmatrix} \begin{bmatrix} 1 & 0 & 0 \\ 0 & (\tfrac{1}{2})^n & 0 \\ 0 & 0 & 0 \end{bmatrix} \begin{bmatrix} 1 & 1 & 1 \\ 0 & -1 & -2 \\ 0 & 0 & 1 \end{bmatrix} \begin{bmatrix} a_0 \\ b_0 \\ c_0 \end{bmatrix}$$

or

$$\mathbf{x}^{(n)} = \begin{bmatrix} a_n \\ b_n \\ c_n \end{bmatrix} = \begin{bmatrix} 1 & 1 - (\tfrac{1}{2})^n & 1 - (\tfrac{1}{2})^{n-1} \\ 0 & (\tfrac{1}{2})^n & (\tfrac{1}{2})^{n-1} \\ 0 & 0 & 0 \end{bmatrix} \begin{bmatrix} a_0 \\ b_0 \\ c_0 \end{bmatrix}$$

$$= \begin{bmatrix} a_0 + b_0 + c_0 - (\tfrac{1}{2})^n b_0 - (\tfrac{1}{2})^{n-1} c_0 \\ (\tfrac{1}{2})^n b_0 + (\tfrac{1}{2})^{n-1} c_0 \\ 0 \end{bmatrix}.$$

Using the fact that $a_0 + b_0 + c_0 = 1$, we thus have

$$a_n = 1 - (\tfrac{1}{2})^n b_0 - (\tfrac{1}{2})^{n-1} c_0$$
$$b_n = (\tfrac{1}{2})^n b_0 + (\tfrac{1}{2})^{n-1} c_0 \qquad n = 1, 2, \ldots \qquad (12.5)$$
$$c_n = 0.$$

These are explicit formulas for the fractions of the three genotypes in the n-th generation of plants in terms of the initial genotype fractions.

Since $(\tfrac{1}{2})^n$ tends to zero as n approaches infinity, it follows from these equations that

$$a_n \to 1$$
$$b_n \to 0$$
$$c_n = 0$$

as n approaches infinity. That is, in the limit all plants in the population will be genotype AA.

EXAMPLE 12.2

Let us modify Example 12.1 so that instead of ferti-
lizing each plant with one of genotype AA, each plant is fertilized
with a plant of its own genotype. Using the same notation as in
Example 12.1, we then find

$$x^{(n)} = M^n x^{(0)}$$

where

$$M = \begin{bmatrix} 1 & \frac{1}{4} & 0 \\ 0 & \frac{1}{2} & 0 \\ 0 & \frac{1}{4} & 1 \end{bmatrix}.$$

The columns of this new matrix M are the same as the columns of
Table 12.1 corresponding to parents with genotypes $AA - AA$, $Aa - Aa$,
and $aa - aa$.

The eigenvalues of M are easily found to be (verify):

$$\lambda_1 = 1, \quad \lambda_2 = 1, \quad \lambda_3 = \frac{1}{2}.$$

The eigenvalue $\lambda_1 = 1$ has multiplicity two and its corresponding eigenspace is two-dimensional. Picking two linearly independent eigenvectors e_1 and e_2 in that eigenspace, and a single eigenvector e_3
for the simple eigenvalue $\lambda_3 = \frac{1}{2}$, we have (verify)

$$e_1 = \begin{bmatrix} 1 \\ 0 \\ 0 \end{bmatrix}, \quad e_2 = \begin{bmatrix} 0 \\ 0 \\ 1 \end{bmatrix}, \quad e_3 = \begin{bmatrix} 1 \\ -2 \\ 1 \end{bmatrix}.$$

The calculations for $x^{(n)}$ are then

$$x^{(n)} = M^n x^{(0)} = PD^n P^{-1} x^{(0)}$$

$$= \begin{bmatrix} 1 & 0 & 1 \\ 0 & 0 & -2 \\ 0 & 1 & 1 \end{bmatrix} \begin{bmatrix} 1 & 0 & 0 \\ 0 & 1 & 0 \\ 0 & 0 & (\frac{1}{2})^n \end{bmatrix} \begin{bmatrix} 1 & \frac{1}{2} & 0 \\ 0 & \frac{1}{2} & 1 \\ 0 & -\frac{1}{2} & 0 \end{bmatrix} \begin{bmatrix} a_0 \\ b_0 \\ c_0 \end{bmatrix}$$

$$= \begin{bmatrix} 1 & \frac{1}{2} - (\frac{1}{2})^{n+1} & 0 \\ 0 & (\frac{1}{2})^n & 0 \\ 0 & \frac{1}{2} - (\frac{1}{2})^{n+1} & 1 \end{bmatrix} \begin{bmatrix} a_0 \\ b_0 \\ c_0 \end{bmatrix}.$$

We thus have

$$a_n = a_0 + [\tfrac{1}{2} - (\tfrac{1}{2})^{n+1}]b_0$$

$$b_n = (\tfrac{1}{2})^n b_0 \qquad\qquad n = 1, 2, \ldots \quad (12.6)$$

$$c_n = c_0 + [\tfrac{1}{2} - (\tfrac{1}{2})^{n+1}]b_0.$$

In the limit, as n tends to infinity, $(\tfrac{1}{2})^n \to 0$ and $(\tfrac{1}{2})^{n+1} \to 0$ so that

$$a_n \to a_0 + \tfrac{1}{2}b_0$$

$$b_n \to 0$$

$$c_n \to c_0 + \tfrac{1}{2}b_0.$$

Thus, fertilization of each plant with one of its own genotype produces a population which in the limit contains only genotypes AA and aa.

AUTOSOMAL RECESSIVE DISEASES

In human populations, there are many genetic diseases governed by autosomal inheritance in which a normal gene A dominates an abnormal gene a. Genotype AA is a normal individual; genotype Aa is a carrier of the disease, though is not afflicted with the disease; and genotype aa is afflicted with the disease. In some cases, such genetic diseases are associated with a particular racial group; e.g., cystic fibrosis (predominant among Caucasians), sickle cell anemia (predominant among Blacks), Cooley's anemia (predominant among people of Mediterranean origin), and Tay-Sachs disease (predominant among Eastern European Jews). It is quite often the case with such diseases that the afflicted individuals die before reaching maturity, so that all afflicted children are the offspring of parents who are both carriers. Suppose a program to identify the carriers of such a disease is carried out, and all of the carriers identified agree not to produce offspring among themselves. In this way, all future children will either have two normal parents (AA - AA matings) or one normal parent and one carrier parent (AA - Aa matings). Consequently, no future children will be afflicted with the disease, though there will still be carriers in future generations. Let us now determine the fraction of carriers in future generations under such a controlled mating program. We set

$$x^{(n)} = \begin{bmatrix} a_n \\ b_n \end{bmatrix}, \qquad n = 1, 2, \ldots$$

where

a_n = fraction of population of genotype AA in n-th
 generation,

b_n = fraction of population of genotype Aa (carriers)
 in n-th generation.

Since each offspring has at least one normal parent, we may consider the controlled mating program as one of continual mating with genotype AA, as in Example 12.1. Thus, the transition of genotype distributions from one generation to the next is governed by the equation

$$\mathbf{x}^{(n)} = M\mathbf{x}^{(n-1)}, \qquad n = 1, 2, \ldots,$$

where

$$M = \begin{bmatrix} 1 & \frac{1}{2} \\ 0 & \frac{1}{2} \end{bmatrix}.$$

Knowing the initial distribution $\mathbf{x}^{(0)}$, the distribution of genotypes in the n-th generation is thus given by

$$\mathbf{x}^{(0)} = M^n \mathbf{x}^{(0)}, \qquad n = 1, 2, \ldots$$

The diagonalization of M is easily carried out (see Exercise 12.4), and leads to

$$\mathbf{x}^{(n)} = PD^n P^{-1} \mathbf{x}^{(0)} = \begin{bmatrix} 1 & 1 \\ 0 & -1 \end{bmatrix} \begin{bmatrix} 1 & 0 \\ 0 & (\frac{1}{2})^n \end{bmatrix} \begin{bmatrix} 1 & 1 \\ 0 & -1 \end{bmatrix} \begin{bmatrix} a_0 \\ b_0 \end{bmatrix}$$

$$= \begin{bmatrix} 1 & 1 - (\frac{1}{2})^n \\ 0 & (\frac{1}{2})^n \end{bmatrix} \begin{bmatrix} a_0 \\ b_0 \end{bmatrix} = \begin{bmatrix} a_0 + b_0 - (\frac{1}{2})^n b_0 \\ (\frac{1}{2})^n b_0 \end{bmatrix}.$$

Since $a_0 + b_0 = 1$, we have

$$a_n = 1 - (\tfrac{1}{2})^n b_0$$
$$b_n = (\tfrac{1}{2})^n b_0 \qquad\qquad n = 1, 2, \ldots \qquad (12.7)$$

As n tends to infinity, we thus have

$$a_n \to 1$$

$$b_n \to 0,$$

so that in the limit there will be no carriers in the population.
From (12.7) we see that

$$b_n = \tfrac{1}{2} b_{n-1} . \qquad n = 1, 2, \ldots \qquad (12.8)$$

That is, the fraction of carriers in each generation is one-half the
fraction of carriers in the preceeding generation. It would be of
interest to also investigate the propagation of carriers under random
mating, when two individuals mate without regard to their genotypes.
Unfortunately, such random mating leads to nonlinear equations, and
the techniques of this chapter are not applicable. However, by
other techniques it can be shown that under random mating Eq. (12.8)
is replaced by

$$b_n = \frac{b_{n-1}}{1 + \tfrac{1}{2} b_{n-1}} \qquad n = 1, 2, \ldots \qquad (12.9)$$

As a numerical example, in the United States, approximately 10% of
Blacks are presently carriers of the gene for sickle cell anemia.
Under a controlled mating program governed by Eq. (12.8), the percentage of carriers can be reduced to 5% in one generation (about
27 years). But under random mating, Eq. (12.9) predicts that 9.5% of
the population will be carriers after one generation (i.e., $b_n = .095$
if $b_{n-1} = .10$). In addition, under controlled mating no offspring
will ever be afflicted with the disease, but with random mating it
can be shown that about 1 in 400 Black children will be born with
sickle cell anemia when 10% of the population are carriers.

X-Linked Inheritance

As mentioned in the introduction, in X-linked inheritance the
male possesses one gene (A or a), and the female possesses two genes
(AA, Aa, or aa). The term "X-linked" is used because such genes are
found on the X-chromosome, of which the male has one and the female
has two. The inheritance of such genes is as follows: a male offspring receives one of his mother's two genes with equal probability,
and a female offspring receives the one gene of her father and one
of her mother's two genes with equal probability. Readers familiar
with basic probability can verify that this type of inheritance
leads to the following table of genotype probabilities.

			Genotypes of Parents (father, mother)					
			(A, AA)	(A, Aa)	(A, aa)	(a, AA)	(a, Aa)	(a, aa)
Offspring	Male	A	1	$\frac{1}{2}$	0	1	$\frac{1}{2}$	0
		a	0	$\frac{1}{2}$	1	0	$\frac{1}{2}$	1
	Female	AA	1	$\frac{1}{2}$	0	0	0	0
		Aa	0	$\frac{1}{2}$	1	1	$\frac{1}{2}$	0
		aa	0	0	0	0	$\frac{1}{2}$	1

Table 12.2

We shall discuss a program of inbreeding in connection with X-linked inheritance. We begin initially with a male and female; select two of their offspring at random, one of each sex, and mate them; select two of the resulting offspring and mate them; and so forth. Such inbreeding is commonly performed with animals. Among humans, such brother-sister marriages were used by the rulers of ancient Egypt to keep the royal line pure.

The original male-female pair can be one of the six types, corresponding to the six columns of Table 12.2:

$$(A, AA), \quad (A, Aa), \quad (A, aa), \quad (a, AA), \quad (a, Aa), \quad (a, aa).$$

The sibling-pairs mated in each successive generation have certain probabilities of being one of these six types. To compute these probabilities, for $n = 0, 1, 2, \ldots$, let us set

a_n = probability sibling-pair mated in n-th generation is type (A, AA)
b_n = " " " " " " " " " (A, Aa)
c_n = " " " " " " " " " (A, aa)
d_n = " " " " " " " " " (a, AA)
e_n = " " " " " " " " " (a, Aa)
f_n = " " " " " " " " " (a, aa)

With these probabilities we form a column vector

$$x^{(n)} = \begin{bmatrix} a_n \\ b_n \\ c_n \\ d_n \\ e_n \\ f_n \end{bmatrix} \qquad n = 0, 1, 2, \ldots$$

From Table 12.2, it is found that

$$x^{(n)} = Mx^{(n-1)}, \qquad n = 1, 2, \ldots \qquad (12.10)$$

where

$$
M =
\begin{array}{c}
\begin{array}{cccccc}
(A,AA) & (A,Aa) & (A,aa) & (a,AA) & (a,Aa) & (a,aa)
\end{array} \\
\left[
\begin{array}{cccccc}
1 & \frac{1}{4} & 0 & 0 & 0 & 0 \\
0 & \frac{1}{4} & 0 & 1 & \frac{1}{4} & 0 \\
0 & 0 & 0 & 0 & \frac{1}{4} & 0 \\
0 & \frac{1}{4} & 0 & 0 & 0 & 0 \\
0 & \frac{1}{4} & 1 & 0 & \frac{1}{4} & 0 \\
0 & 0 & 0 & 0 & \frac{1}{4} & 1
\end{array}
\right]
\begin{array}{l}
(A,AA) \\
(A,Aa) \\
(A,aa) \\
(a,AA) \\
(a,Aa) \\
(a,aa).
\end{array}
\end{array}
$$

For example, suppose in the $(n-1)$st generation the sibling-pair mated is type (A, Aa). Then their male offspring will be either genotype A or a with equal probability, and their female offspring will be either genotype AA or Aa with equal probability. Since one of the male offspring and one of the female offspring are chosen at random for mating, the next sibling-pair will be one of type (A, AA), (A, Aa), (a, AA), (a, Aa), with equal probability. Thus, the second column of M contains "$\frac{1}{4}$" in each of the four rows corresponding to these four sibling-pairs. (See Exercise 12.9 for the remaining columns.)

As in our previous examples, from (12.10) we obtain

$$x^{(n)} = M^n x^{(0)}, \qquad n = 1, 2, \ldots \qquad (12.11)$$

After some lengthly calculations, the eigenvalues and eigenvectors of M turn out to be

$$\lambda_1 = 1, \quad \lambda_2 = 1, \quad \lambda_3 = \tfrac{1}{2}, \quad \lambda_4 = -\tfrac{1}{2}, \quad \lambda_5 = \tfrac{1}{4}(1 + \sqrt{5}), \quad \lambda_6 = \tfrac{1}{4}(1 - \sqrt{5})$$

$$
e_1 = \begin{bmatrix} 1 \\ 0 \\ 0 \\ 0 \\ 0 \\ 0 \end{bmatrix}, \;
e_2 = \begin{bmatrix} 0 \\ 0 \\ 0 \\ 0 \\ 0 \\ 1 \end{bmatrix}, \;
e_3 = \begin{bmatrix} -1 \\ 2 \\ -1 \\ 1 \\ -2 \\ 1 \end{bmatrix}, \;
e_4 = \begin{bmatrix} 1 \\ -6 \\ -3 \\ 3 \\ 6 \\ -1 \end{bmatrix}, \;
e_5 = \begin{bmatrix} \tfrac{1}{4}(-3 - \sqrt{5}) \\ 1 \\ \tfrac{1}{4}(-1 + \sqrt{5}) \\ \tfrac{1}{4}(-1 + \sqrt{5}) \\ 1 \\ \tfrac{1}{4}(-3 - \sqrt{5}) \end{bmatrix}, \;
e_6 = \begin{bmatrix} \tfrac{1}{4}(-3 + \sqrt{5}) \\ 1 \\ \tfrac{1}{4}(-1 - \sqrt{5}) \\ \tfrac{1}{4}(-1 - \sqrt{5}) \\ 1 \\ \tfrac{1}{4}(-3 + \sqrt{5}) \end{bmatrix}.
$$

The diagonalization of M then leads to

$$x^{(n)} = PD^n P^{-1} x^{(0)}, \qquad n = 1, 2, \ldots \qquad (12.12)$$

where

$$P = \begin{bmatrix} 1 & 0 & -1 & 1 & \frac{1}{4}(-3-\sqrt{5}) & \frac{1}{4}(-3+\sqrt{5}) \\ 0 & 0 & 2 & -6 & 1 & 1 \\ 0 & 0 & -1 & -3 & \frac{1}{4}(-1+\sqrt{5}) & \frac{1}{4}(-1-\sqrt{5}) \\ 0 & 0 & 1 & 3 & \frac{1}{4}(-1+\sqrt{5}) & \frac{1}{4}(-1-\sqrt{5}) \\ 0 & 0 & -2 & 6 & 1 & 1 \\ 0 & 1 & 1 & -1 & \frac{1}{4}(-3-\sqrt{5}) & \frac{1}{4}(-3+\sqrt{5}) \end{bmatrix}$$

$$D^n = \begin{bmatrix} 1 & 0 & 0 & 0 & 0 & 0 \\ 0 & 1 & 0 & 0 & 0 & 0 \\ 0 & 0 & (\frac{1}{2})^n & 0 & 0 & 0 \\ 0 & 0 & 0 & (-\frac{1}{2})^n & 0 & 0 \\ 0 & 0 & 0 & 0 & [\frac{1}{4}(1+\sqrt{5})]^n & 0 \\ 0 & 0 & 0 & 0 & 0 & [\frac{1}{4}(1-\sqrt{5})]^n \end{bmatrix}$$

$$P^{-1} = \begin{bmatrix} 1 & 2/3 & 1/3 & 2/3 & 1/3 & 0 \\ 0 & 1/3 & 2/3 & 1/3 & 2/3 & 1 \\ 0 & 1/8 & -1/4 & 1/4 & -1/8 & 0 \\ 0 & -1/24 & -1/12 & 1/12 & 1/24 & 0 \\ 0 & (5+\sqrt{5})/20 & \sqrt{5}/5 & \sqrt{5}/5 & (5+\sqrt{5})/20 & 0 \\ 0 & (5-\sqrt{5})/20 & -\sqrt{5}/5 & -\sqrt{5}/5 & (5-\sqrt{5})/20 & 0 \end{bmatrix}$$

We shall not write out the matrix product in (12.12) since it is rather unwieldy. However, if a specific vector $x^{(0)}$ is given, the calculation for $x^{(n)}$ is not too cumbersome (see Exercise 12.6).

Since the absolute values of the last four diagonal entries of D are less than one, we see that as n tends to infinity

$$D^n \longrightarrow \begin{bmatrix} 1 & 0 & 0 & 0 & 0 & 0 \\ 0 & 1 & 0 & 0 & 0 & 0 \\ 0 & 0 & 0 & 0 & 0 & 0 \\ 0 & 0 & 0 & 0 & 0 & 0 \\ 0 & 0 & 0 & 0 & 0 & 0 \\ 0 & 0 & 0 & 0 & 0 & 0 \end{bmatrix}.$$

And so from Eq. (12.12):

$$x^{(n)} \longrightarrow P \begin{bmatrix} 1 & 0 & 0 & 0 & 0 & 0 \\ 0 & 1 & 0 & 0 & 0 & 0 \\ 0 & 0 & 0 & 0 & 0 & 0 \\ 0 & 0 & 0 & 0 & 0 & 0 \\ 0 & 0 & 0 & 0 & 0 & 0 \\ 0 & 0 & 0 & 0 & 0 & 0 \end{bmatrix} P^{-1} x^{(0)}$$

Performing the matrix multiplication on the right we obtain (verify)

$$x^{(n)} \longrightarrow \begin{bmatrix} a_0 + \frac{2}{3}b_0 + \frac{1}{3}c_0 + \frac{2}{3}d_0 + \frac{1}{3}e_0 \\ 0 \\ 0 \\ 0 \\ 0 \\ f_0 + \frac{1}{3}b_0 + \frac{2}{3}c_0 + \frac{1}{3}d_0 + \frac{2}{3}e_0 \end{bmatrix} \qquad (12.13)$$

That is, in the limit all sibling-pairs will be either type (A, AA) or type (a, aa). For example, if the initial parents are type (A, Aa) (i.e., $b_0 = 1$ and $a_0 = c_0 = d_0 = e_0 = f_0 = 0$), then as n tends to infinity

$$x^{(n)} \longrightarrow \begin{bmatrix} 2/3 \\ 0 \\ 0 \\ 0 \\ 0 \\ 1/3 \end{bmatrix}.$$

Thus, in the limit, there is probability 2/3 that the sibling-pairs will all be (A, AA), and probability 1/3 that they will all be (a, aa).

EXERCISES

12.1 Show that if $M = PDP^{-1}$, then $M^n = PD^nP^{-1}$ for $n = 1, 2, \ldots$.

12.1 In Example 12.1, suppose the plants are always fertilized with a plant of genotype Aa, rather than one of genotype AA. Derive formulas for the fractions of the plants of genotypes AA, Aa, and aa in the n-th generation. Also, find the limiting genotype distribution as n tends to infinity.

12.3 In Example 12.1, suppose the initial plants are fertilized with genotype AA, the first generation is fertilized with genotype Aa, the second generation is fertilized with genotype AA, and this alternating pattern of fertilization is kept up. Find formulas for the fractions of the plants of genotypes AA, Aa, and aa in the n-th generation.

12.4 In the section on autosomal recessive diseases, find the eigenvalues and eigenvectors of the matrix M and verify Eqs. (12.7).

12.5 It is known that carriers of the sickle cell gene have a high resistance to malaria. Consequently, in regions where malaria is common, a balance is reached such that this benevolent aspect of the sickle cell gene balances its malevolent aspect of producing individuals with sickle cell anemia. In certain regions of Africa, this has resulted in an equilibrium situation in which up to 25% of the population are carriers. In the United States, however, malaria is not a problem. Thus, when Blacks from Africa were brought here, only the malevolent aspect of the sickle cell gene remained. The percentage of carriers then began to fall among the Blacks in this country, as described in this chapter, to the point where approximately 10% of Blacks are presently carriers. Using Eq. (12.9), calculate how many generations would be required for the percentage of carriers to fall from 25% to 10% when mating is irrespective of genotype. What would be the corresponding drop for the same number of generations under the controlled mating program determined by Eq. (12.8)?

12.6 In the section on X-linked inheritance, suppose the initial parents are equally likely to be of any of the six possible genotype parents; i.e.,

$$
x^{(0)} = \begin{bmatrix} 1/6 \\ 1/6 \\ 1/6 \\ 1/6 \\ 1/6 \\ 1/6 \end{bmatrix}.
$$

Using Eq. (12.12), calculate $x^{(n)}$, and also calculate the limit of $x^{(n)}$ as n tends to infinity.

12.7 From Eq. (12.13), show that under X-linked inheritance with in-breeding, the probability that the limiting sibling-pairs will be of type (A, AA) is the same as the proportion of A genes in the initial population.

12.8 In X-linked inheritance, suppose none of the females of genotype Aa survive to maturity. Under inbreeding the possible sibling-pairs are then

$$(A, AA), \quad (A, aa), \quad (a, AA), \quad \text{and} \quad (a, aa).$$

Find the transition matrix which describes how the genotype distribution changes in one generation.

12.9 Derive the matrix M in Eq. (12.10) from Table 12.2.

13

Age-Specific Population Growth

The growth in time of a female population divided into age classes is investigated using the Leslie matrix model. The limiting age distribution and growth rate of the population are determined.

PREREQUISITES: Eigenvalues and eigenvectors
Diagonalization of a matrix
Intuitive understanding of limits

INTRODUCTION

One of the most common models of population growth used by demographers is the so-called "Leslie model", developed in the nineteen forties. This model describes the growth of the female portion of a human or animal population. In this model, the females are divided into age classes of equal duration. To be specific, suppose the maximum age attained by any female in the population is L years (or some other time unit) and we divide the population into n age classes. Then each class is L/n years in duration. We label the age classes according to the following table:

Age Class	Age Interval
1	$[0, \ L/n)$
2	$[L/n, \ 2L/n)$
3	$[2L/n, \ 3L/n)$
$\vdots$	$\vdots$
$n-1$	$[(n-2)L/n, \ (n-1)L/n)$
n	$[(n-1)L/n, \ L]$

Suppose we know the number of females in each of the n classes at time $t = 0$. In particular, let there be $x_1^{(0)}$ females in the first class, $x_2^{(0)}$ females in the second class, and so forth. With these n numbers we form a column vector $x^{(0)}$ as follows:

$$x^{(0)} = \begin{bmatrix} x_1^{(0)} \\ x_2^{(0)} \\ \vdots \\ x_n^{(0)} \end{bmatrix}.$$

We call this vector the *initial age distribution vector*.

As time progresses, the number of females within each of the n classes changes because of three biological processes: birth, death, and aging. By describing these three processes quantitatively, we shall see how to project the initial age distribution vector into the future.

The easiest way to study the aging process is to observe the population at discrete times, say $t_0, t_1, t_2, \ldots, t_k, \ldots$. The Leslie model requires that the duration between any two successive observation times be the same as the duration of the age intervals. We therefore set

$$t_0 = 0$$
$$t_1 = L/n$$
$$t_2 = 2L/n$$
$$\vdots$$
$$t_k = kL/n$$
$$\vdots$$

With this assumption, all females in the $(i+1)$st class at time t_{k+1} were in the i-th class at time t_k.

The birth and death processes between two successive observation times may be described by means of the following demographic parameters:

a_i $i = 1, 2, \ldots, n$	The average number of daughters born to a single female during the time she is in the i-th age class.
b_i $i = 1, 2, \ldots, n-1$	The fraction of females in the i-th age class that can be expected to survive and pass into the $(i+1)$st age class.

By their definitions, we have that

(i) $a_i \geq 0$ for $i = 1, 2, \ldots, n$

(ii) $0 < b_i \leq 1$ for $i = 1, 2, \ldots, n-1$.

Notice that we do not allow any b_i to equal zero, since then no females will survive beyond the i-th age class. We also assume that at least one a_i is positive so that some births occur. Any age class for which the corresponding value of a_i is positive is called a *fertile age class*.

We next define the age distribution vector $\mathbf{x}^{(k)}$ at time t_k by

$$\mathbf{x}^{(k)} = \begin{bmatrix} x_1^{(k)} \\ x_2^{(k)} \\ \vdots \\ x_n^{(k)} \end{bmatrix}$$

where $x_i^{(k)}$ is the number of females in the i-th age class at time t_k. Now, at time t_k, the females in the first age class are just those daughters born between times t_{k-1} and t_k. Thus we can write

$$
\begin{Bmatrix}
\text{number of} \\
\text{females} \\
\text{in class 1} \\
\text{at time } t_k
\end{Bmatrix}
=
\begin{Bmatrix}
\text{number of} \\
\text{daughters} \\
\text{born to} \\
\text{females in} \\
\text{class 1} \\
\text{between times} \\
t_{k-1} \text{ and } t_k
\end{Bmatrix}
+
\begin{Bmatrix}
\text{number of} \\
\text{daughters} \\
\text{born to} \\
\text{females in} \\
\text{class 2} \\
\text{between times} \\
t_{k-1} \text{ and } t_k
\end{Bmatrix}
+ \cdots +
\begin{Bmatrix}
\text{number of} \\
\text{daughters} \\
\text{born to} \\
\text{females in} \\
\text{class } n \\
\text{between times} \\
t_{k-1} \text{ and } t_k
\end{Bmatrix}
$$

or mathematically,

$$
x_1^{(k)} = a_1 x_1^{(k-1)} + a_2 x_2^{(k-1)} + \cdots + a_n x_n^{(k-1)} . \tag{13.1}
$$

The number of females in the $(i+1)$st age class ($i = 1, 2, \ldots, n-1$) at time t_k are those females in the i-th class at time t_{k-1} who are still alive at time t_k. Thus,

$$
\begin{Bmatrix}
\text{number of} \\
\text{females in} \\
\text{class } i{+}1 \\
\text{at time } t_k
\end{Bmatrix}
=
\begin{Bmatrix}
\text{fraction of} \\
\text{females in} \\
\text{class } i \\
\text{who survive} \\
\text{and pass into} \\
\text{class } i{+}1
\end{Bmatrix}
\begin{Bmatrix}
\text{number of} \\
\text{females in} \\
\text{class } i \\
\text{at time } t_{k-1}
\end{Bmatrix}
$$

or mathematically,

$$
x_{i+1}^{(k)} = b_i x_i^{(k-1)} , \qquad i = 1, 2, \ldots, n-1 \tag{13.2}
$$

Using matrix notation, Eqs. (13.1) and (13.2) can be written

$$
\begin{bmatrix}
x_1^{(k)} \\
x_2^{(k)} \\
x_3^{(k)} \\
\vdots \\
x_n^{(k)}
\end{bmatrix}
=
\begin{bmatrix}
a_1 & a_2 & a_3 & \cdots & a_{n-1} & a_n \\
b_1 & 0 & 0 & \cdots & 0 & 0 \\
0 & b_2 & 0 & \cdots & 0 & 0 \\
\vdots & \vdots & \vdots & & \vdots & \vdots \\
0 & 0 & 0 & \cdots & b_{n-1} & 0
\end{bmatrix}
\begin{bmatrix}
x_1^{(k-1)} \\
x_2^{(k-1)} \\
x_3^{(k-1)} \\
\vdots \\
x_n^{(k-1)}
\end{bmatrix} ,
$$

or more compactly,

$$x^{(k)} = Lx^{(k-1)}, \qquad k = 1, 2, \ldots \qquad (13.3)$$

where L is the *Leslie matrix*

$$L = \begin{bmatrix} a_1 & a_2 & a_3 & \cdots & a_{n-1} & a_n \\ b_1 & 0 & 0 & \cdots & 0 & 0 \\ 0 & b_2 & 0 & \cdots & 0 & 0 \\ \vdots & \vdots & \vdots & & \vdots & \vdots \\ 0 & 0 & 0 & \cdots & b_{n-1} & 0 \end{bmatrix}. \qquad (13.4)$$

From Eq. (13.3) it follows that

$$\begin{aligned}
x^{(1)} &= Lx^{(0)} \\
x^{(2)} &= Lx^{(1)} = L^2 x^{(0)} \\
x^{(3)} &= Lx^{(2)} = L^3 x^{(0)} \\
&\;\;\vdots \\
x^{(k)} &= Lx^{(k-1)} = L^k x^{(0)}.
\end{aligned} \qquad (13.5)$$

Thus, if we know the initial age distribution $x^{(0)}$ and the Leslie matrix L, we can determine the female age distribution at any later time.

EXAMPLE 13.1 Suppose the oldest age attained by the females in a certain animal population is 15 years, and we divide the population into three equal age classes of durations five years. Let the Leslie matrix for this population be

$$L = \begin{bmatrix} 0 & 4 & 3 \\ \tfrac{1}{2} & 0 & 0 \\ 0 & \tfrac{1}{4} & 0 \end{bmatrix}$$

If there are initially 1,000 females in each of the three age classes, then from Eq. (13.3) we have

$$x^{(0)} = \begin{bmatrix} 1,000 \\ 1,000 \\ 1,000 \end{bmatrix},$$

$$x^{(1)} = Lx^{(0)} = \begin{bmatrix} 0 & 4 & 3 \\ \frac{1}{2} & 0 & 0 \\ 0 & \frac{1}{4} & 0 \end{bmatrix} \begin{bmatrix} 1,000 \\ 1,000 \\ 1,000 \end{bmatrix} = \begin{bmatrix} 7,000 \\ 500 \\ 250 \end{bmatrix},$$

$$x^{(2)} = Lx^{(1)} = \begin{bmatrix} 0 & 4 & 3 \\ \frac{1}{2} & 0 & 0 \\ 0 & \frac{1}{4} & 0 \end{bmatrix} \begin{bmatrix} 7,000 \\ 500 \\ 250 \end{bmatrix} = \begin{bmatrix} 2,750 \\ 3,500 \\ 125 \end{bmatrix},$$

$$x^{(3)} = Lx^{(2)} = \begin{bmatrix} 0 & 4 & 3 \\ \frac{1}{2} & 0 & 0 \\ 0 & \frac{1}{4} & 0 \end{bmatrix} \begin{bmatrix} 2,750 \\ 3,500 \\ 125 \end{bmatrix} = \begin{bmatrix} 14,375 \\ 1,375 \\ 875 \end{bmatrix}.$$

Thus, after 15 years there are 14,375 females between 0 and 5 years of age, 1,375 females between 5 and 10 years of age, and 875 females between 10 and 15 years of age.

LIMITING BEHAVIOR

Although Eq. (13.5) gives the age distribution of the population at any time, it does not immediately give a general picture of the dynamics of the growth process. For this we need to investigate the eigenvalues and eigenvectors of the Leslie matrix. The eigenvalues of L are the roots of its characteristic polynomial. As we ask the reader to verify in Exercise 13.2, this characteristic polynomial is

$$p(\lambda) = |\lambda I - L|$$

$$= \lambda^n - a_1 \lambda^{n-1} - a_2 b_1 \lambda^{n-2} - a_3 b_1 b_2 \lambda^{n-3} - \cdots - a_n b_1 b_2 \cdots b_{n-1}.$$

To analyze the roots of this polynomial, it will be convenient to introduce the function

$$q(\lambda) = \frac{a_1}{\lambda} + \frac{a_2 b_1}{\lambda^2} + \frac{a_3 b_1 b_2}{\lambda^3} + \cdots + \frac{a_n b_1 b_2 \cdots b_{n-1}}{\lambda^n}. \qquad (13.6)$$

Using this function, the characteristic equation $p(\lambda) = 0$ can be written (verify)

$$q(\lambda) = 1 \quad \text{for } \lambda \neq 0. \tag{13.7}$$

Since all of the a_i and b_i are nonnegative, we see that $q(\lambda)$ is monotonically decreasing for λ greater than zero. Furthermore, $q(\lambda)$ has a vertical asymptote at $\lambda = 0$ and approaches zero as $\lambda \to \infty$. Consequently, as Fig. 13.1 indicates, there is a unique λ, say λ_1, such that $q(\lambda) = 1$. That is, the matrix L has a unique positive eigenvalue. It may further be shown (see Exercise 13.3) that λ_1 is simple; i.e., λ_1 has multiplicity one. Although we shall omit the computational details, the reader can verify that an eigenvector corresponding to λ_1, that is, a nonzero vector solution of

$$L\mathbf{x}_1 = \lambda_1 \mathbf{x}_1 ,$$

is

$$\mathbf{x}_1 = \begin{bmatrix} 1 \\ b_1/\lambda_1 \\ b_1 b_2/\lambda_1^2 \\ b_1 b_2 b_3/\lambda_1^3 \\ \vdots \\ b_1 b_2 \cdots b_{n-1}/\lambda_1^{n-1} \end{bmatrix} . \tag{13.8}$$

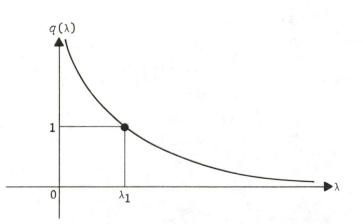

Figure 13.1

Since λ_1 is simple, its corresponding eigenspace has dimension one, and so any eigenvector corresponding to it is some multiple of x_1. Let us summarize these results in the following theorem:

THEOREM 13.1 *A Leslie matrix L has a unique positive eigenvalue λ_1. This eigenvalue is simple and has an eigenvector x_1 all of whose entries are positive.*

We shall now show that the long-term behavior of the age distribution of the population is determined by the positive eigenvalue λ_1 and its eigenvector x_1.

In Exercise 13.9, we ask the reader to prove the following result:

THEOREM 13.2 *If λ_1 is the unique positive eigenvalue of a Leslie matrix L and λ_i is any other real or complex eigenvalue of L, then $|\lambda_i| \leq \lambda_1$.*

Because of Theorem 13.2, λ_1 is called a *dominant eigenvalue* of L. For our purposes we actually need more; namely, that $|\lambda_i| < \lambda_1$ for all other eigenvalues of L. In this case, we say that λ_1 is a *strictly dominant eigenvalue* of L. However, as the following example shows, not all Leslie matrices satisfy this condition.

EXAMPLE 13.2 Let

$$L = \begin{bmatrix} 0 & 0 & 6 \\ 1/2 & 0 & 0 \\ 0 & 1/3 & 0 \end{bmatrix}.$$

Then the characteristic polynomial of L is

$$p(\lambda) = |\lambda I - L| = \lambda^3 - 1.$$

The eigenvalues of L are thus the solutions of $\lambda^3 = 1$; namely,

$$\lambda = 1, \ -\frac{1}{2} + \frac{\sqrt{3}}{2}i, \ -\frac{1}{2} - \frac{\sqrt{3}}{2}i.$$

All three eigenvalues have absolute value one, and so the unique positive eigenvalue $\lambda_1 = 1$ is not strictly dominant. Note that this matrix has the property that $L^3 = I$. This means that for any choice of the initial age distribution $x^{(0)}$, we have

$$x^{(0)} = x^{(3)} = x^{(6)} = \cdots = x^{(3k)} = \cdots .$$

The age distribution vector thus oscillates with a period of three time units. Such oscillations (or *population waves*, as they are called) could not occur if λ_1 were strictly dominant, as we shall see below.

◆━◆━◆

It is beyond the scope of this book to discuss necessary and sufficient conditions for λ_1 to be a strictly dominant eigenvalue. However, we will state the following sufficient condition without proof:

> **THEOREM 13.3** *If two successive entries a_i and a_{i+1} in the first row of a Leslie matrix L are nonzero, then the positive eigenvalue of L is strictly dominant.*

Thus, if the female population has two successive fertile age classes, then its Leslie matrix has a strictly dominant eigenvalue. This is always the case for realistic populations if the duration of the age classes is sufficiently small. Notice that in Example 13.2 there is only one fertile age class (the third), and so the condition of Theorem 13.3 is not satisfied. In what follows, we shall always assume that the condition of Theorem 13.3 is satisfied.

Let us assume that L is diagonalizable. This is not really necessary for the conclusions we shall draw, but it does simplify the arguments. In this case, L has n eigenvalues, $\lambda_1, \lambda_2, \ldots, \lambda_n$, not necessarily distinct, and n linearly independent eigenvectors, $x_1, x_2, \ldots, x_n$, corresponding to them. In this listing we place the strictly dominant eigenvalue λ_1 first. We construct a matrix P whose columns are the eigenvectors of L:

$$P = \left[x_1 \mid x_2 \mid x_3 \mid \cdots \mid x_n \right].$$

The diagonalization of L is then given by the equation

$$
L = P
\begin{bmatrix}
\lambda_1 & 0 & 0 & \cdots & 0 \\
0 & \lambda_2 & 0 & \cdots & 0 \\
\vdots & \vdots & \vdots & & \vdots \\
0 & 0 & 0 & \cdots & \lambda_n
\end{bmatrix}
P^{-1} .
$$

From this it follows that

$$
L^k = P
\begin{bmatrix}
\lambda_1^k & 0 & 0 & \cdots & 0 \\
0 & \lambda_2^k & 0 & \cdots & 0 \\
\vdots & \vdots & \vdots & & \vdots \\
0 & 0 & 0 & \cdots & \lambda_n^k
\end{bmatrix}
P^{-1} ,
$$

for $k = 1, 2, \ldots$. For any initial age distribution vector $x^{(0)}$ we then have

$$
L^k x^{(0)} = P
\begin{bmatrix}
\lambda_1^k & 0 & 0 & \cdots & 0 \\
0 & \lambda_2^k & 0 & \cdots & 0 \\
\vdots & \vdots & \vdots & & \vdots \\
0 & 0 & 0 & \cdots & \lambda_n^k
\end{bmatrix}
P^{-1} x^{(0)}
$$

for $k = 1, 2, \ldots$. Dividing both sides of this equation by λ_1^k and using the fact that $x^{(k)} = L^k x^{(0)}$, we have

$$
\frac{1}{\lambda_1^k} x^{(k)} = P
\begin{bmatrix}
1 & 0 & 0 & \cdots & 0 \\
0 & (\lambda_2/\lambda_1)^k & 0 & \cdots & 0 \\
\vdots & \vdots & \vdots & & \vdots \\
0 & 0 & 0 & \cdots & (\lambda_n/\lambda_1)^k
\end{bmatrix}
P^{-1} x^{(0)} . \tag{13.9}
$$

Since λ_1 is the strictly dominant eigenvalue, $|\lambda_i/\lambda_1| < 1$ for $i = 2, 3, \ldots, n$. It follows that

$$
(\lambda_i/\lambda_1)^k \to 0 \quad \text{as} \quad k \to \infty \quad \text{for} \quad i = 2, 3, \ldots, n.
$$

Using this fact, we may take the limit of both sides of (13.9) to obtain

$$\lim_{k \to \infty} \left\{ \frac{1}{\lambda_1^k} x^{(k)} \right\} = P \begin{bmatrix} 1 & 0 & 0 & \cdots & 0 \\ 0 & 0 & 0 & \cdots & 0 \\ \vdots & \vdots & \vdots & & \vdots \\ 0 & 0 & 0 & \cdots & 0 \end{bmatrix} P^{-1} x^{(0)}. \qquad (13.10)$$

Let us denote the first entry of the column vector $P^{-1} x^{(0)}$ by the constant c. As we ask the reader to show in Exercise 13.4, the righthand side of (13.10 can be written as $c x_1$, where c is a positive constant which depends only on the initial age distribution vector $x^{(0)}$. Thus (13.10) becomes

$$\lim_{k \to \infty} \left\{ \frac{1}{\lambda_1^k} x^{(k)} \right\} = c x_1. \qquad (13.11)$$

Equation (13.11) gives us the approximation

$$x^{(k)} \simeq c \lambda_1^k x_1 \qquad (13.12)$$

for large values of k. From (13.12) we also have

$$x^{(k-1)} \simeq c \lambda_1^{k-1} x_1. \qquad (13.13)$$

Comparing Eqs. (13.12) and (13.13), we see that

$$x^{(k)} \simeq \lambda_1 x^{(k-1)} \qquad (13.14)$$

for large values of k. This means that for large values of time each age distribution vector is a scalar multiple of the preceding age distribution vector, the scalar being the positive eigenvalue of the Leslie matrix. Consequently, the *proportion* of females in each of the age classes becomes constant. As we shall see in the following example, these limiting proportions can be determined from the eigenvector x_1.

EXAMPLE 13.1 (REVISITED)

The Leslie matrix in Example 13.1 was

$$L = \begin{bmatrix} 0 & 4 & 3 \\ \tfrac{1}{2} & 0 & 0 \\ 0 & \tfrac{1}{4} & 0 \end{bmatrix}.$$

Its characteristic polynomial is $p(\lambda) = \lambda^3 - 2\lambda - 3/8$, and the reader can verify that the positive eigenvalue is $\lambda_1 = 3/2$. From (13.8) the corresponding eigenvector x_1 is

$$x_1 = \begin{bmatrix} 1 \\ b_1/\lambda_1 \\ b_1 b_2/\lambda_1^2 \end{bmatrix} = \begin{bmatrix} 1 \\ (1/2)\big/(3/2) \\ (1/2)(1/4)\big/(3/2)^2 \end{bmatrix} = \begin{bmatrix} 1 \\ 1/3 \\ 1/18 \end{bmatrix}.$$

From (13.14) we have

$$x^{(k)} \simeq (3/2) x^{(k-1)}$$

for large values of k. Hence, every five years the number of females in each of the three classes will increase by about 50%, as will the total number of females in the population.
From (13.12) we have

$$x^{(k)} \simeq c(3/2)^k \begin{bmatrix} 1 \\ 1/3 \\ 1/18 \end{bmatrix}.$$

Consequently, eventually the females will be distributed among the three age classes in the ratios $1 : 1/3 : 1/18$. This corresponds to a distribution of 72% of the females in the first age class, 24% of the females in the second age class, and 4% of the females in the third age class.

EXAMPLE 13.3

In this example we shall use birth and death parameters from the year 1965 for Canadian females. Since few women over 50 years of age bear children, we shall restrict ourselves to the portion of the female population between 0 and 50 years of age. The data are for 5-year age classes, so there are a total of ten age classes. Rather than write out the 10×10 Leslie matrix in full, we list the birth and death parameters as follows:

Age Interval	a_i	b_i
[0, 5)	0.00000	0.99651
[5, 10)	0.00024	0.99820
[10, 15)	0.05861	0.99802
[15, 20)	0.28608	0.99729
[20, 25)	0.44791	0.99694
[25, 30)	0.36399	0.99621
[30, 35)	0.22259	0.99460
[35, 40)	0.10457	0.99184
[40, 45)	0.02826	0.98700
[45, 50)	0.00240	—

Using numerical techniques, the positive eigenvalue and corresponding eigenvector turn out to be

$$\lambda_1 = 1.07622 \quad \text{and} \quad x_1 = \begin{bmatrix} 1.00000 \\ 0.92594 \\ 0.85881 \\ 0.79641 \\ 0.73800 \\ 0.68364 \\ 0.63281 \\ 0.58482 \\ 0.53897 \\ 0.49429 \end{bmatrix}.$$

Thus, if Canadian women continued to reproduce and die as they did
in 1965, eventually every five years their numbers would increase by
7.622%. From the eigenvector x_1, we see that, in the limit, for
every 100,000 females between 0 and 5 years of age, there will be
92,594 females between 5 and 10 years of age, 85,881 females between
10 and 15 years of age, and so forth.

◆━◆━◆

Let us look again at Eq. (13.12) which gives the age distribution vector of the population for large times:

$$x^{(k)} \approx c\lambda_1^k x_1 . \tag{13.15}$$

Three cases arise according to the value of the positive eigenvalue λ_1:

(i) The population is eventually increasing if $\lambda_1 > 1$.

(ii) The population is eventually decreasing if $\lambda_1 < 1$.

(iii) The population stabilizes if $\lambda_1 = 1$.

The case $\lambda_1 = 1$ is particularly interesting since it determines a population which has *zero population growth*. For any initial age distribution, the population approaches a limiting age distribution which is some multiple of the eigenvector x_1. From Eqs. (13.6) and (13.7), we see that $\lambda_1 = 1$ is an eigenvalue if and only if

$$a_1 + a_2 b_1 + a_3 b_1 b_2 + \cdots + a_n b_1 b_2 \cdots b_{n-1} = 1. \tag{13.16}$$

The expression

$$R = a_1 + a_2 b_1 + a_3 b_1 b_2 + \cdots + a_n b_1 b_2 \cdots b_{n-1} \tag{13.17}$$

is called the *net reproduction rate* of the population. (See Exercise 13.5 for a demographic interpretation of R.) Thus we can say that a population has zero population growth if and only if its net reproduction rate is one.

EXERCISES

13.1 Suppose a certain animal population is divided into two age classes and has a Leslie matrix

$$L = \begin{bmatrix} 1 & 3/2 \\ 1/2 & 0 \end{bmatrix}.$$

(a) Calculate the positive eigenvalue λ_1 of L and the corresponding eigenvector x_1.

(b) Beginning with the initial age distribution vector

$$x^{(0)} = \begin{bmatrix} 100 \\ 0 \end{bmatrix}$$

calculate $x^{(1)}$, $x^{(2)}$, $x^{(3)}$, $x^{(4)}$, and $x^{(5)}$, rounding off to the nearest integer when necessary.

(c) Calculate $x^{(6)}$ using the exact formula $x^{(6)} = Lx^{(5)}$ and the approximate formula $x^{(6)} \simeq \lambda_1 x^{(5)}$.

13.2 Find the characteristic polynomial of a general Leslie matrix given by Eq. (13.4).

13.3 Show that the positive eigenvalue λ_1 of a Leslie matrix is always simple. Recall that a root λ_0 of a polynomial $q(\lambda)$ is simple if and only if $q'(\lambda_0) \neq 0$.

13.4 Show that the righthand side of Eq. (13.10) is cx_1 where c is the first entry of the column vector $P^{-1}x^{(0)}$.

13.5 Show that the net reproduction rate R, defined by (13.17), can be interpreted as the average number of daughters born to a single female during her expected lifetime.

13.6 Show that a population is eventually decreasing if and only if its net reproduction rate is less than one. Similarly, show that a population is eventually increasing if and only if its net reproduction rate is greater than one.

13.7 Calculate the net reproduction rate of the animal population in Example 13.1.

13.8 (For readers with a hand calculator) Calculate the net reproduction rate of the Canadian female population in Example 13.3.

13.9 (For readers who have had a course in Complex Variables) Prove Theorem 13.2. Hint: write $\lambda_i = re^{i\theta}$, substitute into (13.7), take the real parts of both sides, and show that $r \leq \lambda_1$.

Harvesting of Animal Populations

*The Leslie matrix model of population growth
is used to model the sustainable harvesting
of an animal population. The effect of harvesting different fractions of different age
groups is investigated.*

PREREQUISITES: Chapter 13: Age-Specific Population Growth

INTRODUCTION

In Chapter 13, the Leslie matrix model for the growth of a female population divided into discrete age classes was described. In
this chapter, we shall investigate the effects of harvesting an animal population growing according to such a model. By *harvesting* we
mean the removal of animals from the population. The word "harvesting" is not necessarily a euphemism for "slaughtering"; the animals
may be removed from the population for other purposes.

We shall restrict ourselves to what are called *sustainable
harvesting policies*. By this we mean the following:

Sustainable Harvesting Policy

*A harvesting policy in which an animal population is
periodically harvested is said to be sustainable if the
yield of each harvest is the same and the age distribution of the population remaining after each harvest is
the same.*

Thus, the animal population is not depleted by a sustainable harvesting policy; only the excess growth is exploited.

As in Chapter 13, we shall only discuss the females of the population. If the number of males in each age class is equal to the
number of females — a reasonable assumption for many populations —
then our harvesting policies will also apply to the male portion of
the population.

THE HARVESTING MODEL

Figure 14.1 illustrates the basic idea of the model. We begin
with a population having a particular age distribution. It under-
goes a growth period which will be described by the Leslie matrix.
At the end of the growth period, a certain fraction of each age
class is harvested. The duration of the harvest is to be short in
comparison with the growth period so that any growth or change in
the population during the harvest period may be neglected. Finally,
the population left unharvested is to have the same age distribution
as the original population. This cycle repeats after each harvest,
so that the yield is sustainable.

To describe this harvesting model mathematically, let

$$x = \begin{bmatrix} x_1 \\ x_2 \\ \vdots \\ x_n \end{bmatrix}$$

be the age distribution vector of the population at the beginning of
the growth period. Thus x_i is the number of females in the i-th
class left unharvested. As in Chapter 13, we require that the duration of each age class be identical with the duration of the growth
period. For example, if the population is harvested once a year,
then the population is to be divided into one-year age classes.

Population
before
Growth Period

Population
after
Growth Period

Population
Harvested

Figure 14.1

If L is the Leslie matrix describing the growth of the population, then the vector $L\mathbf{x}$ is the age distribution vector of the population at the end of the growth period, immediately before the periodic harvest. Let h_i, for $i = 1, 2, \ldots, n$, be the fraction of females from the i-th class which is harvested. We use these n numbers to form an $n \times n$ diagonal matrix

$$H = \begin{bmatrix} h_1 & 0 & 0 & \cdots & 0 \\ 0 & h_2 & 0 & \cdots & 0 \\ 0 & 0 & h_3 & \cdots & 0 \\ \vdots & \vdots & \vdots & & \vdots \\ 0 & 0 & 0 & \cdots & h_n \end{bmatrix}$$

which we shall call the *harvesting matrix*. By definition, we have

$$0 \le h_i \le 1 \qquad \text{for } i = 1, 2, \ldots, n.$$

That is, we may harvest none ($h_i = 0$), all ($h_i = 1$), or some proper fraction ($0 < h_i < 1$) of each of the n classes. Since the number of females in the i-th class immediately before each harvest is the i-th entry $(Lx)_i$ of the vector Lx, it can be seen that the i-th entry of the column vector

$$HLx = \begin{bmatrix} h_1 (Lx)_1 \\ h_2 (Lx)_2 \\ \vdots \\ h_n (Lx)_n \end{bmatrix}$$

is the number of females harvested from the i-th class.

From the definition of a sustainable harvesting policy, we have

$$\begin{pmatrix} \text{age distribution} \\ \text{at end of} \\ \text{growth period} \end{pmatrix} - \begin{pmatrix} \text{harvest} \end{pmatrix} = \begin{pmatrix} \text{age distribution} \\ \text{at beginning of} \\ \text{growth period} \end{pmatrix}$$

or mathematically,

$$Lx - HLx = x. \tag{14.1}$$

If we write Eq. (10.1) in the form

$$(I - H)Lx = x \tag{14.2}$$

we see that x must be an eigenvector of the matrix $(I - H)L$ corresponding to the eigenvalue one. As we shall now describe, this places certain restrictions on the values of h_i and x.

Suppose the Leslie matrix of the population is

$$L = \begin{bmatrix} a_1 & a_2 & a_3 & \cdots & a_{n-1} & a_n \\ b_1 & 0 & 0 & \cdots & 0 & 0 \\ 0 & b_2 & 0 & \cdots & 0 & 0 \\ \vdots & \vdots & \vdots & & \vdots & \vdots \\ 0 & 0 & 0 & \cdots & b_{n-1} & 0 \end{bmatrix}.$$

Then the matrix $(I - H)L$ is easily computed:

$$(I - H)L = \begin{bmatrix} (1 - h_1)a_1 & (1 - h_1)a_2 & (1 - h_1)a_3 & \cdots & (1 - h_1)a_{n-1} & (1 - h_1)a_n \\ (1 - h_2)b_1 & 0 & 0 & \cdots & 0 & 0 \\ 0 & (1 - h_3)b_2 & 0 & \cdots & 0 & 0 \\ \vdots & \vdots & \vdots & & \vdots & \vdots \\ 0 & 0 & 0 & \cdots & (1 - h_n)b_{n-1} & 0 \end{bmatrix}.$$

We thus see that $(I - H)L$ is a matrix which has the same mathematical form as a Leslie matrix. In Chapter 13, we showed that a necessary and sufficient condition for a Leslie matrix to have one as an eigenvalue is that its net reproduction rate also be one. (See Eq. (13.16) on page 178.) Calculating the net reproduction rate of $(I - H)L$ and setting it equal to one, we obtain (verify):

$$(1 - h_1)[a_1 + a_2 b_1 (1 - h_2) + a_3 b_1 b_2 (1 - h_2)(1 - h_3) + \cdots$$

$$+ a_n b_1 b_2 \cdots b_{n-1}(1 - h_2)(1 - h_3) \cdots (1 - h_n)] = 1. \quad (14.4)$$

This equation places a restriction on the allowable harvesting fractions. Only those values of $h_1, h_2, \ldots, h_n$ which satisfy Eq. (14.4) and which lie in the interval $[0, 1]$ can produce a sustainable yield. If $h_1, h_2, \ldots, h_n$ do satisfy (14.4), then the matrix $(I - H)L$ has the desired eigenvalue $\lambda_1 = 1$; and, furthermore this eigenvalue has multiplicity one since the positive eigenvalue of a Leslie matrix always has multiplicity one (Theorem 13.1, page 172). This means that there is only one linearly independent eigenvector x satisfying Eq. (14.2). As in Chapter 13 (Eq. (13.8), page 171), we pick the following normalized eigenvector:

$$x_1 = \begin{bmatrix} 1 \\ b_1(1-h_2) \\ b_1 b_2 (1-h_2)(1-h_3) \\ b_1 b_2 b_3 (1-h_2)(1-h_3)(1-h_4) \\ \vdots \\ b_1 b_2 b_3 \cdots b_{n-1}(1-h_2)(1-h_3) \cdots (1-h_n) \end{bmatrix}. \qquad (14.5)$$

Any other solution x of (14.2) is a multiple of x_1. The vector x_1 thus determines the proportion of females within each of the n classes after a harvest under a sustainable harvesting policy. But there is an ambiguity in the total number of females in the population after each harvest. This can be determined by some auxilliary condition, such as an ecological or economic constraint. For example, for a population economically supported by the harvester, the largest population the harvester can afford to raise between harvests would determine the particular constant x_1 is multiplied by to produce the appropriate vector x in (10.2). For a wild population — deer, whales, bears, etc. — the natural habitat of the population would determine how large the total population may be between harvests.

Summarizing our results so far, we see that there is a wide choice in the values of $h_1, h_2, \ldots, h_n$ which will produce a sustainable yield. But once these values are selected, the proportional age distribution of the population after each harvest is uniquely determined by the normalized eigenvector x_1 defined by Eq. (10.5). Let us now consider a few particular harvesting strategies of this type.

UNIFORM HARVESTING

With many populations it is difficult to distinguish or catch animals of specific ages. If animals are caught at random, we may reasonably assume the same fraction of each age class is harvested. Let us therefore set

$$h = h_1 = h_2 = \cdots = h_n.$$

Equation (10.2) then reduces to (verify):

$$Lx = \left(\frac{1}{1-h}\right)x.$$

Hence, $1/(1 - h)$ must be the unique positive eigenvalue λ_1 of the Leslie growth matrix L. That is,

$$\lambda_1 = \frac{1}{1 - h} .$$

Solving for the harvesting fraction h, we obtain

$$h = 1 - 1/\lambda_1. \tag{14.6}$$

The vector x_1, in this case, is the same as the eigenvector of L corresponding to the eigenvalue λ_1. From Chapter 13 (Eq. (13.8), page 115), this is

$$x_1 = \begin{bmatrix} 1 \\ b_1/\lambda_1 \\ b_1 b_2/\lambda_1^2 \\ b_1 b_2 b_3/\lambda_1^3 \\ \vdots \\ b_1 b_2 \cdots b_{n-1}/\lambda_1^{n-1} \end{bmatrix} . \tag{14.7}$$

From Eq. (14.6), we can see that the larger λ_1 is, the larger is the fraction of animals we can harvest without depleting the population. We also notice that we need $\lambda_1 > 1$ in order that the harvesting fraction h lie in the interval $(0, 1]$. This is to be expected since $\lambda_1 > 1$ is the condition that the population be increasing.

EXAMPLE 14.1 For a certain species of domestic sheep in New Zealand with a growth period of one year, the following Leslie matrix was found (G. Caughley, "Parameters for Seasonally Breeding Populations," *Ecology*, Vol. 48, 1967, pages 834 - 839):

$$L = \begin{bmatrix} .000 & .045 & .391 & .472 & .484 & .546 & .543 & .502 & .468 & .459 & .433 & .421 \\ .845 & 0 & 0 & 0 & 0 & 0 & 0 & 0 & 0 & 0 & 0 & 0 \\ 0 & .975 & 0 & 0 & 0 & 0 & 0 & 0 & 0 & 0 & 0 & 0 \\ 0 & 0 & .965 & 0 & 0 & 0 & 0 & 0 & 0 & 0 & 0 & 0 \\ 0 & 0 & 0 & .950 & 0 & 0 & 0 & 0 & 0 & 0 & 0 & 0 \\ 0 & 0 & 0 & 0 & .926 & 0 & 0 & 0 & 0 & 0 & 0 & 0 \\ 0 & 0 & 0 & 0 & 0 & .895 & 0 & 0 & 0 & 0 & 0 & 0 \\ 0 & 0 & 0 & 0 & 0 & 0 & .850 & 0 & 0 & 0 & 0 & 0 \\ 0 & 0 & 0 & 0 & 0 & 0 & 0 & .786 & 0 & 0 & 0 & 0 \\ 0 & 0 & 0 & 0 & 0 & 0 & 0 & 0 & .691 & 0 & 0 & 0 \\ 0 & 0 & 0 & 0 & 0 & 0 & 0 & 0 & 0 & .561 & 0 & 0 \\ 0 & 0 & 0 & 0 & 0 & 0 & 0 & 0 & 0 & 0 & .370 & 0 \end{bmatrix}.$$

The sheep have a lifespan of 12 years so that they are divided into 12 age classes of duration one year each. By the use of numerical techniques, the unique positive eigenvalue of L can be found to be

$$\lambda_1 = 1.221 .$$

From Eq. (10.6), the harvesting fraction h is

$$h = 1 - 1/\lambda_1 = 1 - 1/1.221 = 0.181 .$$

Thus, the uniform harvesting policy is one in which 18.1% of the sheep from each of the 12 age classes is harvested every year. From Eq. (14.7), the age distribution vector of the sheep after each harvest is proportional to

$$x_1 = \begin{bmatrix} 1.000 \\ 0.692 \\ 0.552 \\ 0.436 \\ 0.339 \\ 0.257 \\ 0.189 \\ 0.131 \\ 0.084 \\ 0.048 \\ 0.022 \\ 0.007 \end{bmatrix}. \qquad (14.8)$$

From (14.8) we see that for every 1000 sheep between 0 and 1 years of age which are not harvested, there are 692 sheep between 1 and 2 years of age, 552 sheep between 2 and 3 years of age, and so forth.

Harvesting Only the Youngest Age Class

In some populations, only the youngest females are of any economic value, and so the harvester seeks to harvest only the females from the youngest age class. Accordingly, let us set

$$h_1 = h,$$
$$h_2 = h_3 = \cdots = h_n = 0.$$

Equation (14.4) then reduces to

$$(1 - h)(a_1 + a_2 b_1 + a_3 b_1 b_2 + \cdots + a_n b_1 b_2 \cdots b_{n-1}) = 1,$$

or

$$(1 - h)R = 1$$

where R is the net reproduction rate of the population. (See Eq. (13.17), page 178.) Solving for h, we obtain

$$h = 1 - 1/R. \tag{14.9}$$

We notice from this equation that only if $R > 1$ is a sustainable harvesting policy possible. This is reasonable since only if $R > 1$ is the population increasing. From Eq. (14.5), the age distribution vector after each harvest is proportional to the vector

$$x_1 = \begin{bmatrix} 1 \\ b_1 \\ b_1 b_2 \\ b_1 b_2 b_3 \\ \vdots \\ b_1 b_2 b_3 \cdots b_{n-1} \end{bmatrix}. \tag{14.10}$$

EXAMPLE 14.2 Let us apply this type of sustainable harvesting policy to the sheep population in Example 14.1. For the net reproduction rate of the population, we find:

$$R = a_1 + a_2 b_1 + a_3 b_1 b_2 + \cdots + a_n b_1 b_2 \cdots b_{n-1}$$

$$= (.000) + (.045)(.845) + \cdots + (.421)(.845)(.975)\cdots(.370)$$

$$= 2.513.$$

From Eq. (14.9), the fraction of the first age class harvested is

$$h = 1 - 1/R = 1 - 1/2.513 = .602 \ .$$

From Eq. (14.10), the age distribution of the sheep population after the harvest is proportional to the vector

$$
x_1 = \begin{bmatrix} 1.000 \\ 0.845 \\ (.845)(.975) \\ (.845)(.975)(.965) \\ \\ \vdots \\ \\ (.845)(.975)\cdots(.370) \end{bmatrix} = \begin{bmatrix} 1.000 \\ 0.845 \\ 0.824 \\ 0.795 \\ 0.755 \\ 0.699 \\ 0.626 \\ 0.532 \\ 0.418 \\ 0.289 \\ 0.162 \\ 0.060 \end{bmatrix} . \tag{14.11}
$$

A direct calculation gives us the following (see also Exercise (14.3):

$$
Lx_1 = \begin{bmatrix} 2.513 \\ 0.845 \\ 0.824 \\ 0.795 \\ 0.755 \\ 0.699 \\ 0.626 \\ 0.532 \\ 0.418 \\ 0.289 \\ 0.162 \\ 0.060 \end{bmatrix} . \tag{14.12}
$$

The vector $L\mathbf{x}_1$ is the age distribution vector immediately before the harvest. The total of all entries in $L\mathbf{x}_1$ is 8.518, so that the first entry 2.513 is 29.5% of the total. This means that immediately before each harvest, 29.5% of the population is in the youngest age class. Since 60.2% of this class is harvested, it follows that 17.8% (= 60.2% of 29.5%) of the entire sheep population is harvested each year. This can be compared with the uniform harvesting policy of Example 14.1, in which 18.1% of the sheep population is harvested each year.

OPTIMAL SUSTAINABLE YIELD

We saw in Example 14.1 that a sustainable harvesting policy in which the same fraction of each age class is harvested produces a yield of 18.1% of the sheep population. In Example 14.2, we saw that if only the youngest age class is harvested, the resulting yield is 17.8% of the population. There are many other possible sustainable harvesting policies, and each will provide a generally different yield. It would be of interest to find a sustainable harvesting policy which produces the largest possible yield. Such a policy is called an *optimal sustainable harvesting policy* and the resulting yield is called the *optimal sustainable yield*. However, the determination of the optimal sustainable yield requires Linear Programming theory, and we cannot discuss it in detail in this chapter. (See Chapter 20 for an introduction to Linear Programming theory.) We shall, though, state the following result for reference (J. R. Beddington and D. B. Taylor, "Optimum Age Specific Harvesting of a Population," *Biometrics*, Vol. 29, 1973, pages 801-809):

> ### Optimal Sustainable Yield
>
> *An optimal sustainable harvesting policy is one in which either one or two age classes are harvested. If two age classes are harvested, then the older age class is completely harvested.*

As an illustration, it can be shown using Linear Programming that the optimal sustainable yield of the sheep population is attained when

$$h_1 = 0.522$$
$$h_9 = 1.000$$

$$(14.13)$$

and all other values of h_i are zero. Thus, 52.2% of the sheep between 0 and 1 years of age and all of the sheep between 8 and 9
years of age are harvested. As we ask the reader to show in Exercise 14.2, the resulting optimal sustainable yield is 19.9% of the
population.

EXERCISES

14.1 Let a certain animal population be divided into three one-year
age classes and have as its Leslie matrix

$$L = \begin{bmatrix} 0 & 4 & 3 \\ \frac{1}{2} & 0 & 0 \\ 0 & \frac{1}{4} & 0 \end{bmatrix}.$$

(a) Find the yield and the age distribution vector after each
harvest if the same fraction of each of the three age
classes is harvested every year.

(b) Find the yield and the age distribution vector after each
harvest if only the youngest age class is harvested every
year. Also, find the fraction of the youngest age class
which is harvested.

14.2 For the optimal sustainable harvesting policy described by Eqs.
(14.13), find the vector x_1 which specifies the age distribution of the population after each harvest. Also, calculate the
vector Lx_1 and verify that the optimal sustainable yield is
19.9% of the population.

14.3 If only the first age class of an animal population is harvested, use Eq. (14.10) to show that

$$Lx_1 - x_1 = \begin{bmatrix} R - 1 \\ 0 \\ 0 \\ \vdots \\ 0 \end{bmatrix}$$

where R is the net reproduction rate of the population.

14.4 If only the I-th class of an animal population is to be periodically harvested $(I = 1, 2, \ldots, n)$, find the corresponding harvesting fraction h_I.

14.5 Suppose all of the J-th class and a certain fraction h_I of the
I-th class of an animal population is to be periodically harvested $(1 \leq I < J \leq n)$. Calculate h_I.

Least Squares Fitting to Data

A technique of best fitting a line or other polynomial curve to a set of experimentally determined points in the plane is described.

PREREQUISITES: Linear systems
Matrices
Euclidean space R^n
Column space of a matrix

INTRODUCTION

A common problem in experimental work is to obtain a mathematical relationship $y = f(x)$ between two variables x and y by "fitting" a curve to points in the plane corresponding to various experimentally determined values of x and y, say

$$(x_1, y_1), \quad (x_2, y_2), \quad \ldots, \quad (x_n, y_n) .$$

Based on theoretical considerations, or simply on the pattern of the points, one decides on the general form of the curve $y = f(x)$ to be fitted. Some possibilities are (Fig. 15.1):

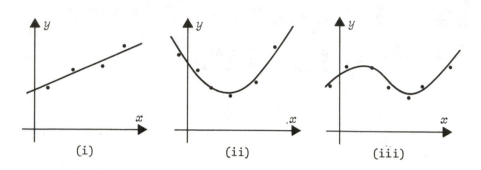

Figure 15.1

(i)   A straight line — $y = a + bx$

(ii) A quadratic polynomial — $y = a + bx + cx^2$

(iii) A cubic polynomial — $y = a + bx + cx^2 + dx^3$.

Because the points are obtained experimentally, there is usually
some "error" in the data making it impossible to find a curve of the
desired form that passes through all of the points. Thus, the idea
is to choose the curve (by determining its coefficients) which
"best" fits the data. We begin with the simplest case: fitting a
straight line to the data points.

LEAST SQUARES FITTING OF A STRAIGHT LINE

Suppose we want to fit a straight line

$$y = a + bx$$

to the experimentally determined points

$$(x_1, y_1), \quad (x_2, y_2), \quad \ldots, \quad (x_n, y_n) .$$

If the data points are collinear, the line would pass through all n
points, and so the unknown coefficients a and b would satisfy

$$y_1 = a + bx_1$$
$$y_2 = a + bx_2$$
$$\vdots$$
$$y_n = a + bx_n.$$

We may write this system in matrix form as

$$\begin{bmatrix} y_1 \\ y_2 \\ \vdots \\ y_n \end{bmatrix} = \begin{bmatrix} 1 & x_1 \\ 1 & x_2 \\ \vdots & \vdots \\ 1 & x_n \end{bmatrix} \begin{bmatrix} a \\ b \end{bmatrix}$$

or more compactly as

$$\mathbf{y} = M\mathbf{v} \qquad (15.1)$$

where

$$\mathbf{y} = \begin{bmatrix} y_1 \\ y_2 \\ \vdots \\ y_n \end{bmatrix}, \qquad M = \begin{bmatrix} 1 & x_1 \\ 1 & x_2 \\ \vdots & \vdots \\ 1 & x_n \end{bmatrix}, \qquad \mathbf{v} = \begin{bmatrix} a \\ b \end{bmatrix}. \qquad (15.2)$$

If the data points are not collinear, it is impossible to find coefficients a and b that satisfy (15.1) exactly; thus no matter how we choose $\mathbf{v}$, the difference

$$\mathbf{y} - M\mathbf{v}$$

between the two sides of (15.1) will not be zero. In this case, our objective will be to find a vector $\mathbf{v}$ which minimizes the Euclidean length of this difference

$$\|\mathbf{y} - M\mathbf{v}\| .$$

If $\mathbf{v}^* = \begin{bmatrix} a^* \\ b^* \end{bmatrix}$ is such a minimizing vector, we call the line $y = a^* + b^*x$ the *least squares straight line fit* to the data.

To see how to find such a vector $\mathbf{v}^*$, refer to Figure 15.2. The vector $\mathbf{y}$ is a fixed vector in R^n. And as $\mathbf{v}$ varies over all possible

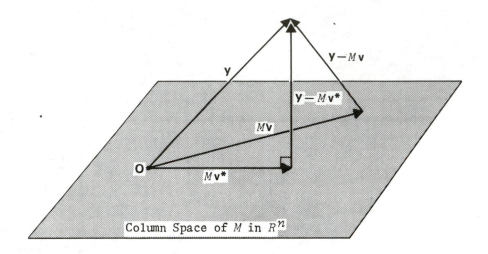

Figure 15.2

values, the vectors $M\mathbf{v}$ form a subspace of R^n — the range space or column space of the matrix M. As suggested by Fig. 15.2, if $\mathbf{y} - M\mathbf{v}$ is to have minimum length it must be orthogonal to the range space. Let $\mathbf{v}^*$ be that vector such that $\mathbf{y} - M\mathbf{v}^*$ is orthogonal to the range space. This means that the Euclidean inner product of $\mathbf{y} - M\mathbf{v}^*$ with $M\mathbf{v}$ is zero for any vector $\mathbf{v}$. Thus,

$$(M\mathbf{v})^t (\mathbf{y} - M\mathbf{v}^*) = 0$$

for all vectors $\mathbf{v}$. Or, since $(M\mathbf{v})^t = \mathbf{v}^t M^t$,

$$\mathbf{v}^t (M^t \mathbf{y} - M^t M\mathbf{v}^*) = 0$$

for all vectors $\mathbf{v}$. But this equation states that the fixed vector $M^t \mathbf{y} - M^t M\mathbf{v}^*$ is orthogonal to every vector $\mathbf{v}$. This is possible only if this fixed vector is the zero vector; that is, if

$$M^t M\mathbf{v}^* = M^t \mathbf{y} . \tag{15.3}$$

Since M is an $n \times 2$ matrix (see Eq. (15.2)), the matrix $M^t M$ is a 2×2 matrix. Thus if $M^t M$ is invertible, (15.3) has the unique solution

$$\mathbf{v}^* = (M^t M)^{-1} M^t \mathbf{y} .$$

REMARK *In Exercise 15.6 we ask the reader to show that M^tM fails to be invertible if and only if the n data points lie on a vertical line in the xy-plane.*

Using this remark, we may summarize our results as follows (where we now write **v** instead of **v*** to simplify the notation):

Given $n > 1$ points (x_1, y_1), (x_2, y_2), $\cdots$, (x_n, y_n), not all lying on a vertical line, the least squares straight line fit to these points

$$y = a + bx,$$

has coefficients given by

$$v = (M^tM)^{-1}M^ty$$

where

$$v = \begin{bmatrix} a \\ b \end{bmatrix}, \qquad M = \begin{bmatrix} 1 & x_1 \\ 1 & x_2 \\ \vdots & \vdots \\ 1 & x_n \end{bmatrix}, \quad and \quad y = \begin{bmatrix} y_1 \\ y_2 \\ \vdots \\ y_n \end{bmatrix}.$$

EXAMPLE 15.1 Find the least squares straight line fit to the four points $(0,1)$, $(1,3)$, $(2,4)$, and $(3,4)$.

SOLUTION We have

$$M = \begin{bmatrix} 1 & 0 \\ 1 & 1 \\ 1 & 2 \\ 1 & 3 \end{bmatrix},$$

$$M^tM = \begin{bmatrix} 4 & 6 \\ 6 & 14 \end{bmatrix},$$

$$(M^tM)^{-1} = \frac{1}{10}\begin{bmatrix} 7 & -3 \\ -3 & 2 \end{bmatrix},$$

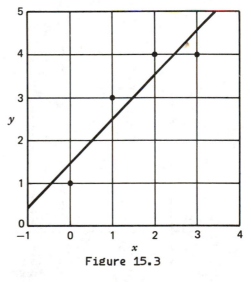

Figure 15.3

$$v = (M^t M)^{-1} M^t y = \frac{1}{10} \begin{bmatrix} 7 & -3 \\ -3 & 2 \end{bmatrix} \begin{bmatrix} 1 & 1 & 1 & 1 \\ 0 & 1 & 2 & 3 \end{bmatrix} \begin{bmatrix} 1 \\ 3 \\ 4 \\ 4 \end{bmatrix} = \begin{bmatrix} 1.5 \\ 1 \end{bmatrix}.$$

And so the desired line is $y = 1.5 + x$.

◆—◆—◆

Figure 15.4 provides an alternate interpretation of least squares fitting of a straight line. We note that $\| y - Mv \|$ and

$$\| y - Mv \|^2 = (y_1 - a - bx_1)^2 + (y_2 - a - bx_2)^2 + \cdots + (y_n - a - bx_n)^2$$

are minimized by the same vector v. From Fig. 15.4 we see that $|y_i - a - bx_i|$ is the vertical distance d_i from the data point (x_i, y_i) to the line $y = a + bx$. If we think of d_i as the "vertical error" at the point (x_i, y_i), then the least squares straight line fit minimizes

$$d_1^2 + d_2^2 + \cdots + d_n^2 ,$$

which is the sum of the squares of the vertical errors.

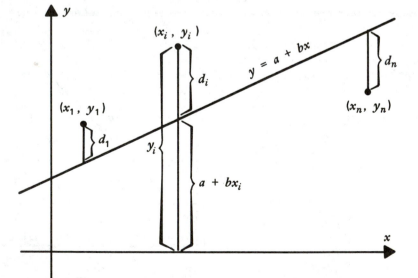

Figure 15.4

EXAMPLE 15.2

Hooke's law in Physics states that the length x of a uniform spring is a linear function of the applied force y. If we write $y = a + bx$, then the coefficient b is called the spring constant. Suppose a particular unstretched spring has a measured length of 6.1 inches (i.e. $x = 6.1$ when $y = 0$). Forces of 2 pounds, 4 pounds, and 6 pounds are then applied to the spring, and the corresponding lengths are found to be 7.6 inches, 8.7 inches, and 10.4 inches. Find the spring constant of this spring.

SOLUTION

We have

$$M = \begin{bmatrix} 1 & 6.1 \\ 1 & 7.6 \\ 1 & 8.7 \\ 1 & 10.4 \end{bmatrix}, \quad y = \begin{bmatrix} 0 \\ 2 \\ 4 \\ 6 \end{bmatrix}$$

and

$$v = \begin{bmatrix} a \\ b \end{bmatrix} = (M^t M)^{-1} M^t y = \begin{bmatrix} -8.6 \\ 1.4 \end{bmatrix}.$$

The spring constant is thus $b = 1.4$ pounds/inch.

x_i	y_i
6.1	0
7.6	2
8.7	4
10.4	6

Figure 15.5

LEAST SQUARES FITTING OF A POLYNOMIAL

The technique described for fitting a straight line to data points generalizes easily to fitting a polynomial of any specified degree to data points. Let us attempt to fit a polynomial of fixed degree m

$$y = a_0 + a_1 x + \cdots + a_m x^m \tag{15.4}$$

to n points

$$(x_1, y_1), \quad (x_2, y_2), \quad \ldots, \quad (x_n, y_n).$$

Substituting these n values of x and y into (15.4) yields the n equations

or, more simply,

$$y = Mv ,$$
(15.5)

where

$$y = \begin{bmatrix} y_1 \\ y_2 \\ \vdots \\ y_n \end{bmatrix}, \qquad M = \begin{bmatrix} 1 & x_1 & x_1^2 & \cdots & x_1^m \\ 1 & x_2 & x_2^2 & \cdots & x_2^m \\ \vdots & \vdots & \vdots & & \vdots \\ 1 & x_n & x_n^2 & \cdots & x_n^m \end{bmatrix}, \qquad v = \begin{bmatrix} a_0 \\ a_1 \\ \vdots \\ a_m \end{bmatrix}.$$

Exactly as before, it can be shown that the least squares solution for v is given by

$$v = (M^t M)^{-1} M^t y,$$
(15.6)

which determines the coefficients of the polynomial. (See Exercise 15.8 for a discussion of the invertibility of the matrix $M^t M$.)

EXAMPLE 15.3

According to Newton's second law of motion, a body near the earth's surface falls vertically downward according to the equation

$$s = s_0 + v_0 t + \tfrac{1}{2} g t^2$$
(15.7)

where

s = vertical displacement downward relative to some fixed point,

s_0 = initial displacement at time $t = 0$,

v_0 = initial velocity at time $t = 0$,

g = acceleration of gravity at earth's surface.

Suppose a laboratory experiment is performed to evaluate g using this equation. A weight is released with unknown initial displacement and velocity and at certain times the distances fallen from some fixed reference point are measured. In particular, suppose that at times $t = .1, .2, .3, .4,$ and $.5$ seconds it is found that the weight has fallen $s = -0.18, 0.31, 1.03, 2.48,$ and 3.73 feet, respectively, from the reference point. Find an approximate value of g using these data.

SOLUTION The mathematical problem is to fit a quadratic curve

$$s = a_0 + a_1 t + a_2 t^2 \tag{15.8}$$

to the five data points:

$$(.1, -0.18), \quad (.2, 0.31), \quad (.3, 1.03), \quad (.4, 2.48), \quad (.5, 3.73).$$

The necessary calculations are

$$M = \begin{bmatrix} 1 & t_1 & t_1^2 \\ 1 & t_2 & t_2^2 \\ 1 & t_3 & t_3^2 \\ 1 & t_4 & t_4^2 \\ 1 & t_5 & t_5^2 \end{bmatrix} = \begin{bmatrix} 1 & .1 & .01 \\ 1 & .2 & .04 \\ 1 & .3 & .09 \\ 1 & .4 & .16 \\ 1 & .5 & .25 \end{bmatrix},$$

$$y = \begin{bmatrix} s_1 \\ s_2 \\ s_3 \\ s_4 \\ s_5 \end{bmatrix} = \begin{bmatrix} -0.18 \\ 0.31 \\ 1.03 \\ 2.48 \\ 3.73 \end{bmatrix},$$

and

$$v = \begin{bmatrix} a_0 \\ a_1 \\ a_2 \end{bmatrix} = (M^t M)^{-1} M^t y = \begin{bmatrix} -0.40 \\ 0.35 \\ 16.1 \end{bmatrix}.$$

From (15.7) and (15.8) we have $a_2 = \frac{1}{2}g$, and so

$$g = 2a_2 = 2(16.1) = 32.2 \text{ feet/second}^2.$$

If desired, we may also evaluate the initial displacement and initial velocity of the weight:

$$s_0 = a_0 = -0.40 \text{ feet},$$

$$v_0 = a_1 = 0.35 \text{ feet/second}.$$

In Fig. 15.6 we have plotted the five data points and the approximating polynomial.

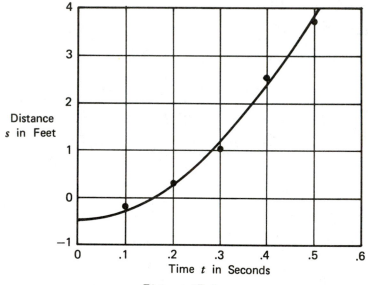

Figure 15.6

EXERCISES

15.1 Find the least squares straight line fit to the three points (0,0), (1,2), and (2,7).

15.2 Find the least squares straight line fit to the four points (0,1), (2,0), (3,1), and (3,2).

15.3 Find the quadratic polynomial which best fits the four points (2,0), (3,-10), (5,-48), and (6,-76).

15.4 Find the cubic polynomial which best fits the five points $(-1,-14)$, $(0,-5)$, $(1,-4)$, $(2,1)$, and $(3,22)$.

15.5 Show that if M is an $m \times n$ matrix with linearly independent columns, then M^tM is an $n \times n$ invertible matrix.

15.6 Show that the matrix M in (15.2) has linearly independent columns if and only if at least two of the numbers $x_1, x_2, \ldots, x_n$ are distinct. Conclude from this and from Exercise 15.5 that the matrix M^tM in (15.3) is invertible if and only if the n data points do not lie on a vertical line in the xy-plane.

15.7 Show that the columns of the $n \times (m+1)$ matrix M in (15.5) are linearly independent if $n > m$ and at least $m+1$ of the numbers $x_1, x_2, \ldots, x_n$ are distinct.

15.8 Using Exercise 15.7, show that a sufficient condition that the matrix M^tM in (15.6) is invertible is that $n > m$ and at least $m+1$ of the numbers $x_1, x_2, \ldots, x_n$ are distinct.

15.9 The owner of a rapidly expanding business finds that for the first five months of the year his sales are $\$4.0, 4.4, 5.2, 6.4$, and 8.0 thousand. He plots these figures on a graph and conjectures that for the rest of the year his sales curve can be approximated by a quadratic polynomial. Find the least squares quadratic polynomial fit to the sales curve and use it to project the sales for the 12th month of the year.

A Least Squares Model for Human Hearing: Fourier Series

The method of least squares approximation is motivated by energy considerations in a model for human hearing. Linear Algebra techniques are used to approximate a function over an interval by a linear combination of sine and cosine terms.

PREREQUISITES: Inner product spaces
Orthogonal projection
$C[a, b]$ (the vector space of continuous
functions on $[a, b]$)

INTRODUCTION

We begin with a brief discussion of the nature of sound and human hearing. Figure 16.1 is a schematic diagram of the ear showing its three main components: the outer ear, middle ear, and inner ear. Sound waves enter the outer ear where they are channeled to the eardrum, causing it to vibrate. Three tiny bones in the middle ear mechanically link the eardrum with the snail-shaped cochlea within the inner ear. These bones pass on the vibrations of the eardrum to a fluid within the cochlea. The cochlea contains thou-

Figure 16.1

sands of minute hairs which oscillate with the fluid. Those near the entrance of the cochlea are stimulated by high frequencies and those near the tip are stimulated by low frequencies. The movements of these hairs activate nerve cells which send signals along various neural pathways to the brain, where the signals are interpreted as sound.

The sound waves themselves are variations in time of the air pressure. For the auditory system, the most elementary type of sound wave is a sinusoidal variation in the air pressure. This type of sound wave stimulates the hairs within the cochlea in such a way that a nerve impulse along a single neural pathway is produced (Fig. 16.2). A sinusoidal sound wave may be described by a function of time

$$q(t) = A_0 + A \sin \omega(t - \delta) \tag{16.1}$$

where $q(t)$ is the atmospheric pressure at the eardrum, A_0 is the normal atmospheric pressure, A is the maximum deviation of the pressure from the normal atmospheric pressure, $\omega/2\pi$ is the frequency of the wave in cycles per second, and δ is the phase angle of the wave. To be perceived as sound, such sinusoidal waves must have frequencies within a certain range. For humans this range is roughly 20 cps to 20,000 cps. Frequencies outside of this range will not stimulate the hairs within the cochlea enough to produce nerve signals.

To a reasonable degree of accuracy, the ear is a linear system. This means that if a complex sound wave is a finite sum of sinusoidal components of different amplitudes, frequencies, and phase angles, say

$$q(t) = A_0 + A_1 \sin \omega_1(t - \delta_1)$$

$$+ A_2 \sin \omega_2(t - \delta_2) + \cdots + A_n \sin \omega_n(t - \delta_n), \tag{16.2}$$

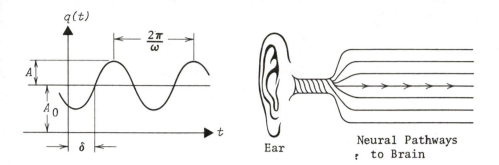

Figure 16.2

then the response of the ear consists of nerve impulses along the same neural pathways that would be stimulated by the individual components (Fig. 16.3).

Let us now consider some periodic sound wave $p(t)$ with period T (i.e., $p(t) \equiv p(t + T)$) which is *not* a finite sum of sinusoidal waves. If we examine the response of the ear to such a periodic wave, we find that it is the same as the response to some wave which is the sum of sinusoidal waves. That is, there is some sound wave $q(t)$ as given by Eq. (16.2) which produces the same response as $p(t)$, even though $p(t)$ and $q(t)$ are different functions of time.

We now want to determine the frequencies, amplitudes, and phase angles of the sinusoidal components of $q(t)$. Since $q(t)$ produces the same response as the periodic wave $p(t)$, it is reasonable to expect that $q(t)$ has the same period T as $p(t)$. This requires that each sinusoidal term in $q(t)$ have period T. Consequently, the frequencies of the sinusoidal components must be integer multiples of

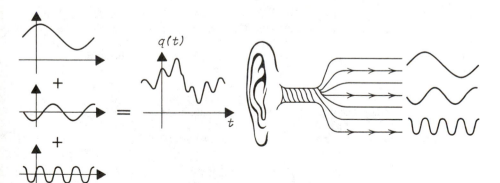

Figure 16.3

the basic frequency $1/T$ of the $p(t)$. That is, the ω_k in Eq. (16.2) must be of the form

$$\omega_k = 2k\pi/T, \qquad k = 1, 2, \ldots$$

But since the ear cannot perceive sinusoidal waves with frequencies greater than 20,000 cps, we may omit those values of k for which $\omega_k/2\pi = k/T$ is greater than 20,000. Thus, $q(t)$ is of the form

$$q(t) = A_0 + A_1 \sin \frac{2\pi}{T}(t - \delta_1) + \cdots + A_n \sin \frac{2n\pi}{T}(t - \delta_n) \qquad (16.3)$$

where n is the largest integer such that n/T is not greater than 20,000.

We now turn our attention to the values of the amplitudes A_0, $A_1, \ldots, A_n$ and the phase angles $\delta_1, \delta_2, \ldots, \delta_n$ which appear in Eq. (12.3). There is some criterion by which the auditory system "picks" these values so that $q(t)$ produces the same response as $p(t)$. To examine this criterion, let us set

$$e(t) = p(t) - q(t).$$

If we consider $q(t)$ as an approximation to $p(t)$, then $e(t)$ is the error in this approximation; an error which the ear cannot perceive. In terms of $e(t)$, the criterion for the determination of the amplitudes and the phase angles is that the quantity

$$\int_0^T [e(t)]^2 dt$$

be as small as possible. We cannot go into the physiological reasons for this, but we can remark that this expression is proportional to the *acoustic energy* of the error wave $e(t)$ over one period. In other words, it is the energy of the difference between the two sound waves $p(t)$ and $q(t)$ which determine whether the ear perceives any difference between them. If this energy is as small as possible, then the two waves produce the same sensation of sound.

In general, suppose we want to approximate a function $p(t)$ by another function $q(t)$ from a certain class. If the criterion of approximation is the smallness of the integral

$$\int_0^T [p(t) - q(t)]^2 dt, \qquad (16.4)$$

then we call $q(t)$ the *least squares approximation* to $p(t)$ over the interval $[0, T]$. The integral in (16.4) is called the *least squares error*, or the *mean squares error*, of the approximation. This criterion arises quite naturally in a wide variety of engineering and scientific approximation problems. Besides the acoustics problem just discussed, some other examples are

1. Let $T(x)$ be the temperature distribution in a uniform rod lying along the x-axis from $x = 0$ to $x = \ell$ (Fig. 16.4). The thermal energy in the rod is proportional to the integral

$$\int_0^\ell [T(x)]^2 dx.$$

The closeness of an approximation $q(x)$ to $T(x)$ can be judged according to the thermal energy of the difference of the two temperature distributions. That energy is proportional to

Figure 16.4

$$\int_0^\ell [T(x) - q(x)]^2 dx,$$

which is a least squares criterion.

2. Let $E(t)$ be a periodic voltage across a resistor in an electrical circuit. The electrical energy transferred to the resistor during one period T is proportional to

$$\int_0^T [E(t)]^2 dt.$$

If $q(t)$ has the same period as $E(t)$ and is to be an approximation to $E(t)$, then the criterion

Figure 16.5

of closeness might be taken as the energy of the difference voltage. This is proportional to

$$\int_0^T [E(t) - q(t)]^2 dt,$$

which is again a least squares criterion.

3. Let $y(x)$ be the vertical displacement of a uniform flexible string whose equilibrium position is along the x-axis from $x = 0$ to $x = \ell$. The elastic potential energy of the string is proportional to

$$\int_0^\ell [y(x)]^2 dx.$$

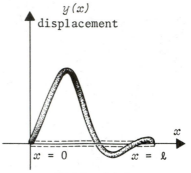

$y(x)$ displacement

$x = 0$ $x = \ell$

Figure 16.6

If $q(x)$ is to be an approximation to the displacement, then as before, the energy integral

$$\int_0^\ell [y(x) - q(x)]^2 dx$$

determines a least squares criterion for the closeness of the approximation.

Least squares approximation is also used in situations where there is no a priori justification for its use, such as for approximating business cycles, population growth curves, sales curves, and so forth. It is used in these cases because of its mathematical simplicity. In general, if no other error criterion is immediately apparent for an approximation problem, the least squares criterion is the one most often chosen.

In the next section we develop the mathematical theory of the least squares approximation of a function by a linear combination of sinusoidal functions.

GENERAL THEORY

Let $f(t)$ be a given continuous function defined over an interval of the t-axis. We first consider the case when the interval is

$[0, 2\pi]$ and later consider the general case when the interval is $[0, T]$ for arbitrary T. Analogous to Eq. (16.3) with $T = 2\pi$, we desire to approximate $f(t)$ by a function of the form

$$g(t) = A_0 + A_1\sin(t - \delta_1) + A_2\sin 2(t - \delta_2) + \cdots + A_n\sin n(t - \delta_n)$$

for some fixed integer n. Since $\sin k(t - \delta_k)$ is expressible as a linear combination of $\sin kt$ and $\cos kt$, we can write $g(t)$ in the alternate form

$$g(t) = c_0 + c_1\cos t + c_2\cos 2t + \cdots + c_n\cos nt$$

$$+ d_1\sin t + d_2\sin 2t + \cdots + d_n\sin nt . \qquad (16.5)$$

Such a function is called a *trigonometric polynomial of order n*. Our problem is to find values of $c_0, c_1, \ldots, c_n, d_1, \ldots, d_n$ such that $g(t)$ is the least squares approximation to $f(t)$ over the interval $[0, 2\pi]$. That is, the coefficients are to be chosen so that the least squares error

$$\int_0^{2\pi} [f(t) - g(t)]^2 dt \qquad (16.6)$$

is as small as possible.

Since the integral in Eq. (16.6) is a function of the $2n + 1$ coefficients $c_0, c_1, \ldots, c_n, d_1, \ldots, d_n$, it is possible to use calculus to find the minimum value of the least squares error and the corresponding values of these $2n + 1$ coefficients. However, an approach using Linear Algebra will give us greater insight into the nature of the approximation process. Moreover, the method we discuss can be applied to many least square problems besides those in this text. We will need the following three facts:

1. The function $f(t)$ we are attempting to approximate may be viewed as a vector in the vector space $C[0, 2\pi]$ — the space of all continuous functions on $[0, 2\pi]$.

2. Since the approximating function $g(t)$ is a linear combination of $1, \cos t, \ldots, \cos nt, \sin t, \ldots, \sin nt$, we may view $g(t)$ as a vector in the subspace W of $C[0, 2\pi]$ spanned by these $2n + 1$ vectors.

3. Since

$$\|f - g\| = \sqrt{\int_0^{2\pi} [f(t) - g(t)]^2 dt} \qquad (16.7)$$

is the distance between $f(t)$ and $g(t)$ in the norm generated by the inner product

$$<u, v> = \int_0^{2\pi} u(t)v(t)\, dt, \qquad (16.8)$$

the least squares error

$$\int_0^{2\pi} [f(t) - g(t)]^2 dt \qquad (16.9)$$

represents the square of the distance $\|f - g\|$.

In light of these remarks, the problem of finding a trigonometric polynomial $g(t)$ which minimizes the least squares error given by (16.9) is equivalent to the problem of finding a vector g in the subspace W which minimizes the distance $\|f - g\|$. The latter problem can be solved by use of the following theorem from the theory of inner product spaces (Fig. 16.7).

THEOREM 16.1 *Let f be a vector in an inner product space and let W be a finite-dimensional subspace. Then the vector g in W which minimizes the distance $\|f - g\|$ is $\text{proj}_W f$, the orthogonal projection of f onto W. If the vectors*

$$g_0, g_1, \ldots, g_m$$

form an orthonormal basis for W, then $\text{proj}_W f$ is given by

$$\text{proj}_W f = <f, g_0>g_0 + <f, g_1>g_1 + \cdots + <f, g_m>g_m \qquad (16.10)$$

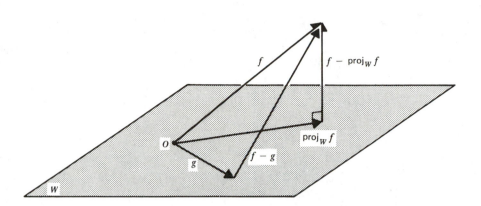

Figure 16.7

To apply this theorem, we must first find an orthonormal basis for the subspace W spanned by the $2n + 1$ vectors $1, \cos t, \cos 2t, \ldots,$ $\cos nt, \sin t, \sin 2t, \ldots, \sin nt$. A direct calculation (see Exercise 16.6) verifies that these $2n + 1$ vectors are orthogonal relative to the inner product (16.8). Consequently, we need only divide each one of these vectors by its length to generate an orthonormal basis for W. The result is (see Exercise 16.7):

$$g_0 = \frac{1}{\sqrt{2\pi}}, \, g_1 = \frac{1}{\sqrt{\pi}} \cos t, \ldots, \quad g_n = \frac{1}{\sqrt{\pi}} \cos nt,$$

$$g_{n+1} = \frac{1}{\sqrt{\pi}} \sin t, \ldots, g_{2n} = \frac{1}{\sqrt{\pi}} \sin nt.$$

The orthogonal projection of f onto W is then given by (16.10):

$$\text{proj}_W f = \langle f, g_0 \rangle \frac{1}{\sqrt{2\pi}} + \langle f, g_1 \rangle \frac{1}{\sqrt{\pi}} \cos t + \cdots + \langle f, g_n \rangle \frac{1}{\sqrt{\pi}} \cos nt$$

$$+ \langle f, g_{n+1} \rangle \frac{1}{\sqrt{\pi}} \sin t + \cdots + \langle f, g_{2n} \rangle \frac{1}{\sqrt{\pi}} \sin nt. \quad (16.11)$$

To simplify our notation, let us define

$$a_0 = 2 \langle f, g_0 \rangle \frac{1}{\sqrt{2\pi}} = \frac{1}{\pi} \int_0^{2\pi} f(t) dt \,,$$

$$a_k = \langle f, g_k \rangle \frac{1}{\sqrt{\pi}} = \frac{1}{\pi} \int_0^{2\pi} f(t) \cos kt \, dt, \qquad k = 1, 2, \ldots, n$$

$$b_k = \langle f, g_{n+k} \rangle \frac{1}{\sqrt{\pi}} = \frac{1}{\pi} \int_0^{2\pi} f(t) \sin kt \, dt, \qquad k = 1, 2, \ldots, n.$$

Equation (16.11) can then be written as

$$\text{proj}_W f = \tfrac{1}{2} a_0 + a_1 \cos t + \cdots + a_n \cos nt + b_1 \sin t + \cdots + b_n \sin nt.$$

In summary, we have the following result:

THEOREM 16.2 *If $f(t)$ is continuous on $[0, 2\pi]$, the tri-gonometric function $g(t)$ of the form*

$$g(t) = \tfrac{1}{2} a_0 + a_1 \cos t + \cdots + a_n \cos nt + b_1 \sin t + \cdots + b_n \sin nt$$

which minimizes the least squares error

$$\int_0^{2\pi} [f(t) - g(t)]^2 dt$$

has coefficients

$$a_k = \frac{1}{\pi} \int_0^{2\pi} f(t) \cos kt \, dt, \qquad k = 0, 1, 2, \ldots, n$$

$$b_k = \frac{1}{\pi} \int_0^{2\pi} f(t) \sin kt \, dt, \qquad k = 1, 2, \ldots, n.$$

If the original function $f(t)$ is defined over the interval $[0, T]$ instead of $[0, 2\pi]$, a change of scale will yield the following result (see Exercise 16.8):

THEOREM 16.3 *If $f(t)$ is continuous on $[0, T]$, the trigonometric function $g(t)$ of the form*

$$g(t) = \tfrac{1}{2}a_0 + a_1\cos\frac{2\pi}{T}t + \cdots + a_n\cos\frac{2n\pi}{T}t$$

$$+ b_1\sin\frac{2\pi}{T}t + \cdots + b_n\sin\frac{2n\pi}{T}t$$

which minimizes the least squares error

$$\int_0^T [f(t) - g(t)]^2\,dt$$

has coefficients

$$a_k = \frac{2}{T}\int_0^T f(t)\cos\frac{2k\pi t}{T}\,dt, \qquad k = 0, 1, 2, \ldots, n$$

$$b_k = \frac{2}{T}\int_0^T f(t)\sin\frac{2k\pi t}{T}\,dt, \qquad k = 1, 2, \ldots, n.$$

EXAMPLE 16.1

Let a sound wave $p(t)$ have a saw-tooth pattern with a basic frequency of 5000 cps (Fig. 16.8). Units have been chosen so that the normal atmospheric pressure is at the zero level, and the maximum amplitude of the wave is A. The basic period of the wave is $T = 1/5000 = .0002$ seconds. From $t = 0$ to $t = T$, $p(t)$ has the equation

$$p(t) = \frac{2A}{T}\left(\frac{T}{2} - t\right).$$

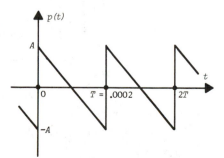

Figure 16.8

Theorem 16.3 then yields the following (verify):

$$a_0 = \frac{2}{T} \int_0^T p(t)\, dt = \frac{2}{T} \int_0^T \frac{2A}{T}\left(\frac{T}{2} - t\right) dt = 0,$$

$$a_k = \frac{2}{T} \int_0^T p(t)\cos\frac{2k\pi t}{T}\, dt = \frac{2}{T} \int_0^T \frac{2A}{T}\left(\frac{T}{2} - t\right)\cos\frac{2k\pi t}{T}\, dt = 0, \quad k = 1, 2, \ldots$$

$$b_k = \frac{2}{T} \int_0^T p(t)\sin\frac{2k\pi t}{T}\, dt = \frac{2}{T} \int_0^T \frac{2A}{T}\left(\frac{T}{2} - t\right)\sin\frac{2k\pi t}{T}\, dt = \frac{2A}{k\pi},$$
$$k = 1, 2, \ldots$$

Let us investigate how the sound wave $p(t)$ is perceived by the human ear. We notice that $4/T = 20,000$ cps, so that we need only go up to $k = 4$ in the above formulas. The least squares approximation to $p(t)$ is then

$$q(t) = \frac{2A}{\pi}\left[\sin\frac{2\pi}{T}t + \frac{1}{2}\sin\frac{4\pi}{T}t + \frac{1}{3}\sin\frac{6\pi}{T}t + \frac{1}{4}\sin\frac{8\pi}{T}t\right].$$

The four sinusoidal terms have frequencies of 5,000, 10,000, 15,000, and 20,000 cps, respectively. In Fig. 16.9 we have plotted $p(t)$ and $q(t)$ over one period. Although $q(t)$ is not a very good point-by-point approximation to $p(t)$, to the ear, both $p(t)$ and $q(t)$ produce the same sensation of sound.

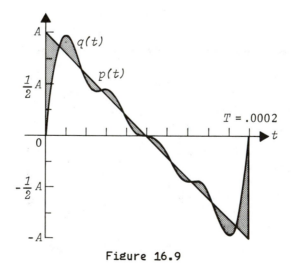

Figure 16.9

As might be expected, the least squares approximation becomes better as the number of terms in the approximating trigonometric polynomial becomes larger. In more advanced courses, it is shown that the least squares error tends to zero as n approaches infinity. For a function $f(t)$ defined over the interval $[0, 2\pi]$, this limiting approximation is denoted by

$$f(t) = \tfrac{1}{2}a_0 + \sum_{k=1}^{\infty}(a_k\cos kt + b_k\sin kt) \tag{16.12}$$

and is called the *Fourier series* of $f(t)$ on the interval $[0, 2\pi]$. The equality in this equation denotes an equality between the two sides of the equation considered as vectors in $C[0, 2\pi]$. To be precise, Eq. (16.12) denotes the fact that the quantity

$$\int_0^{2\pi}\left[f(t) - \tfrac{1}{2}a_0 - \sum_{k=1}^{n}(a_k\cos kt + b_k\sin kt)\right]^2 dt$$

tends to zero as n approaches infinity. Whether the Fourier series of $f(t)$ converges to $f(t)$ for each t is another question, and a more difficult one. For most continuous functions encountered in applications, the Fourier series does indeed converge to its corresponding function for each value of t.

EXERCISES

16.1 Find the trigonometric polynomial of order three which is the least squares approximation to the function $f(t) = (t - \pi)^2$ over the interval $[0, 2\pi]$.

16.2 Find the trigonometric polynomial of order four which is the least squares approximation to the function $f(t) = t^2$ over the interval $[0, T]$.

16.3 Find the trigonometric polynomial of order four which is the least squares approximation to the function $f(t)$ over the interval $[0, 2\pi]$ where

$$f(t) = \begin{cases} \sin t & 0 \le t \le \pi \\ 0 & \pi < t \le 2\pi \end{cases}.$$

16.4 Find the trigonometric polynomial of arbitrary order n which is the least squares approximation to the function $f(t) = \sin \frac{1}{2} t$ over the interval $[0, 2\pi]$.

16.5 Find the trigonometric polynomial of arbitrary order n which is the least squares approximation to the function $f(t)$ over the interval $[0, T]$ where

$$f(t) = \begin{cases} t & 0 \leq t \leq \frac{1}{2}T \\ T - t & \frac{1}{2}T < t \leq T \end{cases}.$$

16.6 Show that the $2n + 1$ functions

$$1, \cos t, \cos 2t, \ldots, \cos nt, \sin t, \sin 2t, \ldots, \sin nt$$

are orthogonal over the interval $[0, 2\pi]$ relative to the inner product $\langle u, v \rangle$ defined by Eq. (16.8).

16.7 For the distance formula defined in (16.7), show that

(a) $\|1\| = \sqrt{2\pi}$

(b) $\|\cos kt\| = \sqrt{\pi}$ for $k = 1, 2, \ldots$

(c) $\|\sin kt\| = \sqrt{\pi}$ for $k = 1, 2, \ldots$

16.8 If $f(t)$ is defined and continuous on the interval $[0, T]$, show that $f(T\tau/2\pi)$ is defined and continuous for τ in the interval $[0, 2\pi]$. Use this fact to show how Theorem 16.3 follows from Theorem 16.2.

Cubic Spline Interpolation

An artist's drafting aid is used as a physical model for the mathematical problem of finding a curve that passes through specified points in the plane. The parameters of the curve are determined by solving a linear system.

PREREQUISITES: Linear systems
Matrix algebra
Differential calculus

INTRODUCTION

Fitting a curve through specified points in a plane is a common problem encountered in analyzing experimental data, in ascertaining the relationships between variables, and in design work. In Figure 17.1, seven points in the *xy*-plane are displayed and in Figure 17.2 a smooth curve has been drawn that passes through them. A curve that passes through a set of points in the plane is said to *interpolate* those points and the curve is called an *interpolating* curve for those points. The interpolating curve in Figure 17.2 was drawn with the aid of a *drafting spline* (Fig. 17.3). This drafting aid consists of a thin, flexible strip

221

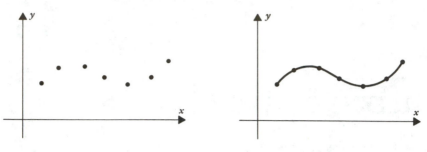

Figure 17.1   Figure 17.2

of wood or other material that is bent to pass through the points
to be interpolated. Attached weights hold the spline in position
while the artist draws the interpolating curve. The drafting
spline will serve as the physical model for a mathematical theory
of interpolation that we shall discuss in this chapter.

STATEMENT OF THE PROBLEM

Suppose we are given n points in the xy-plane,

$$(x_1, y_1), (x_2, y_2), \ldots , (x_n, y_n),$$

that we wish to interpolate with a "well-behaved" curve (Fig. 17.4).
For convenience, we take the points to be equally spaced in the
x-direction, although our results can easily be extended to the case
of unequally spaced points. If we let the common distance between
the x-coordinates of the points be h, then we have

$$x_2 - x_1 = x_3 - x_2 = \cdots = x_n - x_{n-1} = h.$$

Let $y = S(x)$, $x_1 \leq x \leq x_n$, denote the interpolating curve that we
seek. We shall assume that this curve describes the displacement
of a drafting spline that interpolates the n points when the
weights holding down the spline are situated precisely at the n
points. It is known from linear beam theory that the fourth
derivative of the displacement of a beam is zero along any interval

Figure 17.3

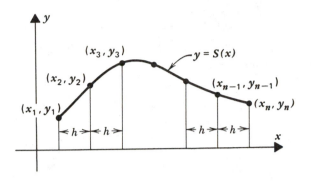

Figure 17.4

of the x-axis that contains no external forces acting on the beam.
If we treat our drafting spline as a thin beam and realize that
the only external forces acting on it arise from the weights at
the n specified points, then it follows that

$$S^{(iv)}(x) \equiv 0 \qquad (17.1)$$

for values of x lying in the $n-1$ open intervals

$$(x_1, x_2), \ (x_2, x_3), \ \ldots, \ (x_{n-1}, x_n)$$

between the n points.

We shall also need the result from linear beam theory which
states that for a beam acted upon only by external forces the
displacement must have two continuous derivatives. In the case
of the interpolating curve $y = S(x)$ constructed by the drafting
spline, this means that $S(x)$, $S'(x)$, and $S''(x)$ must be continuous
for $x_1 \le x \le x_n$.

The condition that $S''(x)$ be continuous is what causes a
drafting spline to produce a pleasing curve, since it results in
a curve with a continuous *curvature*. The eye can perceive sudden
changes in curvature, that is, discontinuities in $S''(x)$, but sudden
changes in higher derivatives are not discernible. Thus, the
condition that $S''(x)$ be continuous is the minimal prerequisite for
the interpolating curve to be perceptible as a single smooth

curve, rather than as a series of separate curves pieced together.

To determine the mathematical form of $S(x)$, we observe that since $S^{(iv)}(x) \equiv 0$ in the intervals between the n specified points, it follows by integrating this equation four times that $S(x)$ must be a *cubic polynomial* in x in each such interval. In general, however, $S(x)$ will be a different cubic polynomial in each interval, and so $S(x)$ must have the form

$$S(x) = \begin{cases} S_1(x), & x_1 \leq x \leq x_2 \\ S_2(x), & x_2 \leq x \leq x_3 \\ \quad \vdots \\ S_{n-1}(x), & x_{n-1} \leq x \leq x_n, \end{cases} \qquad (17.2)$$

where $S_1(x)$, $S_2(x)$, ..., $S_{n-1}(x)$ are cubic polynomials. For convenience, we will write these in the form

$$S_1(x) = a_1(x - x_1)^3 + b_1(x - x_1)^2 + c_1(x - x_1) + d_1, \qquad x_1 \leq x \leq x_2$$

$$S_2(x) = a_2(x - x_2)^3 + b_2(x - x_2)^2 + c_2(x - x_2) + d_2, \qquad x_2 \leq x \leq x_3$$

$$\vdots \qquad (17.3)$$

$$S_{n-1}(x) = a_{n-1}(x - x_{n-1})^3 + b_{n-1}(x - x_{n-1})^2$$

$$+ c_{n-1}(x - x_{n-1}) + d_{n-1}, \qquad x_{n-1} \leq x \leq x_n$$

The a_i's, b_i's, c_i's, and d_i's constitute a total of $4n-4$ coefficients that we must determine to completely specify $S(x)$. If we choose these coefficients so that $S(x)$ interpolates the n specified points in the plane and $S(x)$, $S'(x)$, and $S''(x)$ are continous, then the resulting interpolating curve is called a *cubic spline*.

DERIVATION OF THE FORMULA OF A CUBIC SPLINE

From Eqs. (17.2) and (17.3), we have

$$S(x) = S_1(x) = a_1(x - x_1)^3 + b_1(x - x_1)^2 + c_1(x - x_1) + d_1, \qquad x_1 \le x \le x_2$$

$$S(x) = S_2(x) = a_2(x - x_2)^3 + b_2(x - x_2)^2 + c_2(x - x_2) + d_2, \qquad x_2 \le x \le x_3$$

$$\vdots \qquad \vdots \qquad\qquad\qquad (17.4)$$

$$S(x) = S_{n-1}(x) = a_{n-1}(x - x_{n-1})^3 + b_{n-1}(x - x_{n-1})^2$$

$$+ c_{n-1}(x - x_{n-1}) + d_{n-1}, \qquad x_{n-1} \le x \le x_n$$

and so

$$S'(x) = S_1'(x) = 3a_1(x - x_1)^2 + 2b_1(x - x_1) + c_1, \qquad x_1 \le x \le x_2$$

$$S'(x) = S_2'(x) = 3a_2(x - x_2)^2 + 2b_2(x - x_2) + c_2, \qquad x_2 \le x \le x_3$$

$$\vdots \qquad \vdots$$

$$S'(x) = S_{n-1}'(x) = 3a_{n-1}(x - x_{n-1})^2$$

$$+ 2b_{n-1}(x - x_{n-1}) + c_{n-1}, \qquad x_{n-1} \le x \le x_n$$

$$(17.5)$$

and

$$S''(x) = S_1''(x) = 6a_1(x - x_1) + 2b_1, \qquad x_1 \le x \le x_2$$

$$S''(x) = S_2''(x) = 6a_2(x - x_2) + 2b_2, \qquad x_2 \le x \le x_3$$

$$\vdots \qquad \vdots$$

$$S''(x) = S_{n-1}''(x) = 6a_{n-1}(x - x_{n-1}) + 2b_{n-1}, \qquad x_{n-1} \le x \le x_n .$$

$$(17.6)$$

We will now use these equations and the four properties of cubic splines stated below to express the unknown coefficients a_i, b_i, c_i, d_i, $i = 1, 2, \ldots, n-1$, in terms of the known coordinates y_1, y_2, $\ldots$, y_n.

(1) *S(x) interpolates the points* (x_i, y_i), $i = 1, 2, \ldots, n$.

Since $S(x)$ interpolates the points (x_i, y_i), $i = 1, 2, \ldots, n$, we have

$$S(x_1) = y_1, \; S(x_2) = y_2, \; \ldots, \; S(x_n) = y_n. \qquad (17.7)$$

From the first $n-1$ of these equations and (17.4) we obtain

$$d_1 = y_1$$
$$d_2 = y_2$$
$$\vdots$$
$$d_{n-1} = y_{n-1}. \qquad (17.8)$$

From the last equation in (17.7), the last equation in (17.4), and the fact that $x_n - x_{n-1} = h$, we obtain

$$a_{n-1}h^3 + b_{n-1}h^2 + c_{n-1}h + d_{n-1} = y_n. \qquad (17.9)$$

(2) *S(x) is continuous on* $[x_1, x_n]$.

Since $S(x)$ is continuous for $x_1 \leq x \leq x_n$, it follows that at each point x_i in the set $x_2, x_3, \ldots, x_{n-1}$ we must have

$$S_{i-1}(x_i) = S_i(x_i), \; i = 2, 3, \ldots, n-1 \qquad (17.10)$$

for otherwise the graphs of $S_{i-1}(x)$ and $S_i(x)$ would not join together to form a continuous curve at x_i. Using the interpolating property $S_i(x_i) = y_i$, it follows from (17.10) that $S_{i-1}(x_i) = y_i$, $i = 2, 3, \ldots, n-1$, or from

(17.4) that

$$a_1 h^3 + b_1 h^2 + c_1 h + d_1 = y_2$$

$$a_2 h^3 + b_2 h^2 + c_2 h + d_2 = y_3$$

$$\vdots$$

$$a_{n-2} h^3 + b_{n-2} h^2 + c_{n-2} h + d_{n-2} = y_{n-1}. \qquad (17.11)$$

(3) $S'(x)$ is continuous on $[x_1, x_n]$,

Since $S'(x)$ is continuous for $x_1 \le x \le x_n$, it follows that

$$S'_{i-1}(x_i) = S'_i(x_i), \quad i = 2, 3, \ldots, n-1,$$

or from (17.5)

$$3a_1 h^2 + 2b_1 h + c_1 = c_2$$

$$3a_2 h^2 + 2b_2 h + c_2 = c_3$$

$$\vdots$$

$$3a_{n-2} h^2 + 2b_{n-2} h + c_{n-2} = c_{n-1}. \qquad (17.12)$$

(4) $S''(x)$ is continous on $[x_1, x_2]$.

Since $S''(x)$ is continuous for $x_1 \le x \le x_n$, it follows that

$$S''_{i-1}(x_i) = S''(x_i), \quad i = 2, 3, \ldots, n-1$$

or from (17.6)

$$6a_1 h + 2b_1 = 2b_2$$

$$6a_2 h + 2b_2 = 2b_3$$

$$\vdots$$

$$6a_{n-2} h + 2b_{n-2} = 2b_{n-1} \qquad (17.13)$$

Equations (17.8), (17.9), (17.11), (17.12), and (17.13) constitute a system of $4n-6$ linear equations in the $4n-4$ unknown
coefficients a_i, b_i, c_i, d_i, $i = 1, 2, \ldots, n-1$. Consequently, we
need two more equations to determine these coefficients uniquely.
Before obtaining these additional equations, however, we will
simplify our existing system by expressing the unknowns a_i, b_i,
c_i, and d_i in terms of new unknown quantities,

$$M_1 = S''(x_1), \ M_2 = S''(x_2), \ \ldots, \ M_n = S''(x_n)$$

and the known quantities

$$y_1, \ y_2, \ \ldots, \ y_n.$$

For example, from (17.6) it follows that

$$M_1 = 2b_1$$

$$M_2 = 2b_2$$

$$\vdots$$

$$M_{n-1} = 2b_{n-1}$$

and so

$$b_1 = \tfrac{1}{2}M_1, \ b_2 = \tfrac{1}{2}M_2, \ \ldots, \ b_{n-1} = \tfrac{1}{2}M_{n-1}.$$

Moreover, we have already seen, (17.8), that

$$d_1 = y_1, \ d_2 = y_2, \ \ldots, \ d_{n-1} = y_{n-1}.$$

We leave it as an exercise for the reader to derive the expressions for the a_i's and c_i's in terms of the M_i's and y_i's. The
final result is as follows:

THEOREM 17.1 *Given n points (x_1, y_1), (x_2, y_2), ... , (x_n, y_n) with $x_{i+1} - x_i = h$, $i = 1, 2, ... , n-1$, the cubic spline*

$$S(x) = \begin{cases} a_1(x - x_1)^3 + b_1(x - x_1)^2 + c_1(x - x_1) + d_1, \\ \qquad\qquad\qquad\qquad\qquad\qquad x_1 \le x \le x_2 \\[2ex] a_2(x - x_2)^3 + b_2(x - x_2)^2 + c_2(x - x_2) + d_2, \\ \qquad\qquad\qquad\qquad\qquad\qquad x_2 \le x \le x_3 \\[2ex] \quad\vdots \qquad\qquad\quad \vdots \\[1ex] a_{n-1}(x - x_{n-1})^3 + b_{n-1}(x - x_{n-1})^2 \\ \quad + c_{n-1}(x - x_{n-1}) + d_{n-1}, \qquad x_{n-1} \le x \le x_n \end{cases}$$

that interpolates these points has coefficients given by

$$a_i = (M_{i+1} - M_i)/6h.$$
$$b_i = M_i/2,$$
$$c_i = (y_{i+1} - y_i)/h - (M_{i+1} + 2M_i)h/6,$$
$$d_i = y_i, \qquad\qquad\qquad\qquad\qquad (17.14)$$

for $i = 1, 2, ... , n-1$, where $M_i = S''(x_i)$, $i = 1, 2, ... , n$.

From this result, we see that the quantities $M_1, M_2, ... , M_n$ uniquely determine the cubic spline. To find these quantities we substitute the expressions for a_i, b_i, and c_i given in (17.14) into (17.12). After some algebraic simplification, we obtain

$$M_1 + 4M_2 + M_3 = 6(y_1 - 2y_2 + y_3)/h^2$$

$$M_2 + 4M_3 + M_4 = 6(y_2 - 2y_3 + y_4)/h^2 \qquad (17.15)$$

$$\vdots$$

$$M_{n-2} + 4M_{n-1} + M_n = 6(y_{n-2} - 2y_{n-1} + y_n)/h^2,$$

or in matrix form:

$$
\begin{bmatrix}
1 & 4 & 1 & 0 & \cdots & 0 & 0 & 0 & 0 \\
0 & 1 & 4 & 1 & \cdots & 0 & 0 & 0 & 0 \\
0 & 0 & 1 & 4 & \cdots & 0 & 0 & 0 & 0 \\
\vdots & \vdots & \vdots & \vdots & & \vdots & \vdots & \vdots & \vdots \\
0 & 0 & 0 & 0 & \cdots & 4 & 1 & 0 & 0 \\
0 & 0 & 0 & 0 & \cdots & 1 & 4 & 1 & 0 \\
0 & 0 & 0 & 0 & \cdots & 0 & 1 & 4 & 1
\end{bmatrix}
\begin{bmatrix}
M_1 \\ M_2 \\ M_3 \\ M_4 \\ \vdots \\ M_{n-3} \\ M_{n-2} \\ M_{n-1} \\ M_n
\end{bmatrix}
= \frac{6}{h^2}
\begin{bmatrix}
y_1 - 2y_2 + y_3 \\
y_2 - 2y_3 + y_4 \\
y_3 - 2y_4 + y_5 \\
\vdots \\
y_{n-4} - 2y_{n-3} + y_{n-2} \\
y_{n-3} - 2y_{n-2} + y_{n-1} \\
y_{n-2} - 2y_{n-1} + y_n
\end{bmatrix}
$$

This is a linear system of $n-2$ equations for the n unknowns M_1, $M_2, \ldots, M_n$. Thus, we still need two additional equations to determine $M_1, M_2, \ldots, M_n$ uniquely. The reason for this is that there are infinitely many cubic splines that interpolate the given points, so that we simply do not have enough conditions to determine a unique cubic spline passing through the points. Below, we discuss three possible ways of specifying the two additional conditions required to obtain a unique cubic spline through the points. (The exercises discuss two more.) They are summarized in Table 17.1.

THE NATURAL SPLINE

The two simplest mathematical conditions we can impose are

$$M_1 = M_n = 0.$$

These conditions together with (17.15) result in an $n \times n$ linear system, for $M_1, M_2, \ldots, M_n$, which can be written in matrix form as

$$
\begin{bmatrix}
1 & 0 & 0 & 0 & \cdots & 0 & 0 & 0 \\
1 & 4 & 1 & 0 & \cdots & 0 & 0 & 0 \\
0 & 1 & 4 & 1 & \cdots & 0 & 0 & 0 \\
\vdots & \vdots & \vdots & \vdots & & \vdots & \vdots & \vdots \\
0 & 0 & 0 & 0 & \cdots & 1 & 4 & 1 \\
0 & 0 & 0 & 0 & \cdots & 0 & 0 & 1
\end{bmatrix}
\begin{bmatrix}
M_1 \\ M_2 \\ M_3 \\ \vdots \\ M_{n-1} \\ M_n
\end{bmatrix}
= \frac{6}{h^2}
\begin{bmatrix}
0 \\
y_1 - 2y_2 + y_3 \\
y_2 - 2y_3 + y_4 \\
\vdots \\
y_{n-2} - 2y_{n-1} + y_n \\
0
\end{bmatrix}
$$

For numerical calculations it is more convenient to eliminate M_1 and M_n from this system and write

$$
\begin{bmatrix}
4 & 1 & 0 & 0 & \cdots & 0 & 0 & 0 \\
1 & 4 & 1 & 0 & \cdots & 0 & 0 & 0 \\
0 & 1 & 4 & 1 & \cdots & 0 & 0 & 0 \\
\vdots & \vdots & \vdots & \vdots & & \vdots & \vdots & \vdots \\
0 & 0 & 0 & 0 & \cdots & 1 & 4 & 1 \\
0 & 0 & 0 & 0 & \cdots & 0 & 1 & 4
\end{bmatrix}
\begin{bmatrix}
M_2 \\ M_3 \\ M_4 \\ \vdots \\ M_{n-2} \\ M_{n-1}
\end{bmatrix}
= \frac{6}{h^2}
\begin{bmatrix}
y_1 - 2y_2 + y_3 \\
y_2 - 2y_3 + y_4 \\
y_3 - 2y_4 + y_5 \\
\vdots \\
y_{n-3} - 2y_{n-2} + y_{n-1} \\
y_{n-2} - 2y_{n-1} + y_n
\end{bmatrix}
$$

$$(17.16)$$

together with

$$M_1 = 0 \tag{17.17}$$

$$M_n = 0 . \tag{17.18}$$

Thus, the $(n-2) \times (n-2)$ linear system can be solved for the $n-2$ coefficients $M_2, M_3, \ldots, M_{n-1}$ and M_1 and M_n can be determined by the last two (trivial) equations.

Physically, the natural spline results when the ends of a drafting spline extend freely beyond the interpolating points without constraint. The end portions of the spline outside the interpolating points will fall on straight line paths causing $S''(x)$ to vanish at the end points x_1 and x_n and resulting in the mathematical conditions $M_1 = M_n = 0$.

The natural spline tends to flatten the interpolating curve at the endpoints, which may be undesirable. Of course, if it is required that $S''(x)$ vanish at the endpoints, then the natural spline must be used.

THE PARABOLIC RUNOUT SPLINE

The two additional constraints imposed for this type of spline are

$$M_1 = M_2 \tag{17.19}$$

$$M_n = M_{n-1}. \tag{17.20}$$

If we use these two equations to eliminate M_1 and M_n from (17.15) we obtain the $(n-2) \times (n-2)$ linear system

$$
\begin{bmatrix}
5 & 1 & 0 & 0 & \cdots & 0 & 0 & 0 \\
1 & 4 & 1 & 0 & \cdots & 0 & 0 & 0 \\
0 & 1 & 4 & 1 & \cdots & 0 & 0 & 0 \\
\vdots & \vdots & \vdots & \vdots & & \vdots & \vdots & \vdots \\
0 & 0 & 0 & 0 & \cdots & 1 & 4 & 1 \\
0 & 0 & 0 & 0 & \cdots & 0 & 1 & 5
\end{bmatrix}
\begin{bmatrix}
M_2 \\ M_3 \\ M_4 \\ \vdots \\ M_{n-2} \\ M_{n-1}
\end{bmatrix}
= \frac{6}{h^2}
\begin{bmatrix}
y_1 - 2y_2 + y_3 \\
y_2 - 2y_3 + y_4 \\
y_3 - 2y_4 + y_5 \\
\vdots \\
y_{n-3} - 2y_{n-2} + y_{n-1} \\
y_{n-2} - 2y_{n-1} + y_n
\end{bmatrix}
\tag{17.21}
$$

for $M_2, M_3, \ldots, M_n$. Once these $n-2$ values have been determined, then M_1 and M_n are determined from (17.19) and (17.20).

From (17.14) we see that $M_1 = M_2$ implies that $a_1 = 0$, and $M_n = M_{n-1}$ implies that $a_{n-1} = 0$. Thus, from (17.3) there are no cubic terms in the formula for the spline over the end intervals $[x_1, x_2]$ and $[x_{n-1}, x_n]$. Hence, as the name suggests, the parabolic runout spline reduces to a parabolic curve over these end intervals.

THE CUBIC RUNOUT SPLINE

For this type of spline, we impose the two additional conditions

$$M_1 = 2M_2 - M_3 \qquad (17.22)$$

$$M_n = 2M_{n-1} - M_{n-2} . \qquad (17.23)$$

Using these two equations to eliminate M_1 and M_n from Eqs. (17.15) results in the following $(n \times 2) \times (n-2)$ linear system for M_2, M_3, ..., M_{n-1} :

$$
\begin{bmatrix}
6 & 0 & 0 & 0 & \cdots & 0 & 0 & 0 \\
1 & 4 & 1 & 0 & \cdots & 0 & 0 & 0 \\
0 & 1 & 4 & 1 & \cdots & 0 & 0 & 0 \\
\vdots & \vdots & \vdots & \vdots & & \vdots & \vdots & \vdots \\
0 & 0 & 0 & 0 & \cdots & 1 & 4 & 1 \\
0 & 0 & 0 & 0 & \cdots & 0 & 0 & 6
\end{bmatrix}
\begin{bmatrix}
M_2 \\ M_3 \\ M_4 \\ \vdots \\ M_{n-2} \\ M_{n-1}
\end{bmatrix}
= \frac{6}{h^2}
\begin{bmatrix}
y_1 - 2y_2 + y_3 \\
y_2 - 2y_3 + y_4 \\
y_3 - 2y_4 + y_5 \\
\vdots \\
y_{n-3} - 2y_{n-2} + y_{n-1} \\
y_{n-2} - 2y_{n-1} + y_n
\end{bmatrix} .
$$

$$(17.24)$$

After solving this linear system for M_2, M_3, ..., M_{n-1}, Eqs. (17.22) and (17.23) can be used to determine M_1 and M_n.

If we rewrite Eq. (17.22) as

$$M_2 - M_1 = M_3 - M_2 .$$

it then follows from (17.14) that $a_1 = a_2$. Since $S'''(x) = 6a_1$ on $[x_1, x_2]$ and $S'''(x) = 6a_2$ on $[x_2, x_3]$, we see that $S'''(x)$ is constant over the entire interval $[x_1, x_3]$. Consequently, $S(x)$ consists of a *single* cubic curve over the interval $[x_1, x_3]$ rather than two different cubic curves pieced together at x_2. (To see this integrate $S'''(x)$ three times.) A similar analysis shows that $S(x)$ consists of a single cubic curve over the last two intervals. Whereas the natural spline tends to produce an interpolating curve that is flat at the endpoints, the cubic runout spline has the opposite tendency: it produces a curve with pronounced curvature at the endpoints. If neither behavior is desired, the parabolic runout spline is a reasonable compromise.

Table 17.1

		Boundary conditions	Matrix equation
NATURAL SPLINE	The second derivative of the spline is zero at the endpoints.	$M_1 = 0$ $M_n = 0$	$$\begin{bmatrix} 4 & 1 & 0 & \cdots & 0 & 0 \\ 1 & 4 & 1 & \cdots & 0 & 0 \\ & & & \cdots & & \\ 0 & 0 & 0 & \cdots & 4 & 1 \\ 0 & 0 & 0 & \cdots & 1 & 4 \end{bmatrix} \begin{bmatrix} M_2 \\ M_3 \\ \vdots \\ M_{n-2} \\ M_{n-1} \end{bmatrix} = \frac{6}{h^2} \begin{bmatrix} y_1 - 2y_2 + y_3 \\ y_2 - 2y_3 + y_4 \\ \cdots \\ y_{n-2} - 2y_{n-1} + y_n \end{bmatrix}$$
PARABOLIC RUNOUT SPLINE	The spline reduces to a parabolic curve on the first and last intervals.	$M_1 = M_2$ $M_n = M_{n-1}$	$$\begin{bmatrix} 5 & 1 & 0 & \cdots & 0 & 0 \\ 1 & 4 & 1 & \cdots & 0 & 0 \\ & & & \cdots & & \\ 0 & 0 & 0 & \cdots & 4 & 1 \\ 0 & 0 & 0 & \cdots & 1 & 5 \end{bmatrix} \begin{bmatrix} M_2 \\ M_3 \\ \vdots \\ M_{n-2} \\ M_{n-1} \end{bmatrix} = \frac{6}{h^2} \begin{bmatrix} y_1 - 2y_2 + y_3 \\ y_2 - 2y_3 + y_4 \\ \cdots \\ y_{n-2} - 2y_{n-1} + y_n \end{bmatrix}$$
CUBIC RUNOUT SPLINE	The spline is a single cubic curve on the first two and last two intervals.	$M_1 = 2M_2 - M_3$ $M_n = 2M_{n-1} - M_{n-2}$	$$\begin{bmatrix} 6 & 0 & 0 & \cdots & 0 & 0 \\ 1 & 4 & 1 & \cdots & 0 & 0 \\ & & & \cdots & & \\ 0 & 0 & 0 & \cdots & 4 & 1 \\ 0 & 0 & 0 & \cdots & 0 & 6 \end{bmatrix} \begin{bmatrix} M_2 \\ M_3 \\ \vdots \\ M_{n-2} \\ M_{n-1} \end{bmatrix} = \frac{6}{h^2} \begin{bmatrix} y_1 - 2y_2 + y_3 \\ y_2 - 2y_3 + y_4 \\ \cdots \\ y_{n-2} - 2y_{n-1} + y_n \end{bmatrix}$$

EXAMPLE

The density of water is well known to reach a maximum at a temperature slightly above freezing. Table 17.2, taken from the *Handbook of Chemistry and Physics*, gives the density of water in grams per cubic centimeter for five equally spaced temperatures from -10°C to 30°C. We will interpolate these five tempera-ture-density measurements with a parabolic runout spline and attempt to find the maximum density of water in this range by finding the maximum

Temperature °C	Density gm/cm^3
-10	.99815
0	.99987
10	.99973
20	.99823
30	.99567

Table 17.2

value on this cubic spline.   In the exercises, we ask the reader to perform similar calculations using a natural spline and a cubic runout spline to interpolate the data points.

Set

$$x_1 = -10, \quad y_1 = .99815$$

$$x_2 = \quad 0, \quad y_2 = .99987$$

$$x_3 = \quad 10, \quad y_3 = .99973$$

$$x_4 = \quad 20, \quad y_4 = .99823$$

$$x_5 = \quad 30, \quad y_5 = .99567 .$$

Then

$$6[y_1 - 2y_2 + y_3]/h^2 = -.0001116$$

$$6[y_2 - 2y_3 + y_4]/h^2 = -.0000816$$

$$6[y_3 - 2y_4 + y_5]/h^2 = -.0000636$$

and the linear system (17.21) for the parabolic runout spline becomes

$$\begin{bmatrix} 5 & 1 & 0 \\ 1 & 4 & 1 \\ 0 & 1 & 5 \end{bmatrix} \begin{bmatrix} M_2 \\ M_3 \\ M_4 \end{bmatrix} = \begin{bmatrix} -.0001116 \\ -.0000816 \\ -.0000636 \end{bmatrix} .$$

Solving this system yields

$$M_2 = -.00001973$$

$$M_3 = -.00001293$$

$$M_4 = -.00001013$$

From (17.19) and (17.20) we have

$$M_1 = M_2 = -.00001973$$

$$M_5 = M_4 = -.00001013$$

Solving for the a_i's, b_i's, c_i's, and d_i's from Eqs. (17.14), we obtain the following expression for the interpolating parabolic runout spline:

$$S(x) = \begin{cases} -.00000987\,(x+10)^2 + .0002707\,(x+10) + .99815, & -10 \leq x \leq 0 \\ .000000113\,(x-0)^3 - .00000987\,(x-0)^2 + .0000733\,(x-0) + .99987, & 0 \leq x \leq 10 \\ .000000047\,(x-10)^3 - .00000647\,(x-10)^2 - .0000900\,(x-10) + .99973, & 10 \leq x \leq 20 \\ -.00000507\,(x-20)^2 - .0002054\,(x-20) + .99823, & 20 \leq x \leq 30 \end{cases}$$

This spline is plotted in Figure 17.5. From that figure we see that the maximum is attained in the interval [0, 10]. To find this maximum, we set $S'(x)$ equal to zero in the interval [0, 10]:

$$S'(x) = .000000339x^2 - .0000197x + .0000733 = 0 \,.$$

To three significant digits the root of this quadratic in the interval [0, 10] is $x = 3.99$, and for this value of x, $S(3.99) = 1.00001$. Thus, according to our interpolated estimate, the maximum density of water is 1.00001 gm/cm^3 attained at 3.99°C. This agrees well with the experimental maximum density of 1.00000 gm/cm^3 attained at 3.98°C. (In the original metric system, the gram was *defined* as the mass of one cubic centimeter of water at its maximum density.)

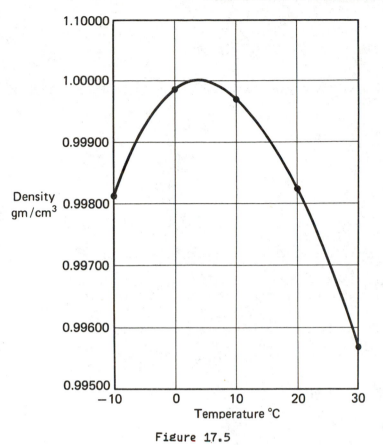

Figure 17.5

CLOSING REMARKS

In addition to producing excellent interpolating curves, cubic splines and their generalizations are useful for numerical integration and differentiation, for the numerical solution of differential and integral equations, and in optimization theory. Readers desiring to pursue this topic further might find the following references useful:

[1]   Ahlberg, J.H., Nilson, E.N., and Walsh, J.L., *The Theory of Splines and their Applications*, Academic Press, 1967

[2]   de Boor, C., *A Practical Guide to Splines*, Springer-Verlag, 1978.

[3]   Rice, J.R., *Numerical Methods, Software, and Analysis*, McGraw-Hill, 1983.

EXERCISES

17.1 Derive the expressions for a_i and c_i in Eqs. (17.14) of Theorem 17.1.

17.2 The six points

$$(\ \ 0, \ .00000)$$
$$(\ .2, \ .19867)$$
$$(\ .4, \ .38942)$$
$$(\ .6, \ .56464)$$
$$(\ .8, \ .71736)$$
$$(1.0, \ .84147)$$

lie on the graph of $y = \sin x$, where x is in radians.

(a) Find the parabolic runout spline that interpolates these six points for $.4 \le x \le .6$. Maintain five decimal place accuracy in your calculations.

(b) Calculate $S(.5)$ for the spline you found in part (a). What is the percentage error of $S(.5)$ with respect to the "exact" value of $\sin(.5) = .47943$?

17.3 The following five points

$$(0, 1)$$
$$(1, 7)$$
$$(2, 27)$$
$$(3, 79)$$
$$(4, 181)$$

lie on a single cubic curve.

(a) Which of the three types of cubic splines (natural, parabolic runout, or cubic runout) would agree exactly with the single cubic curve on which the five points lie?

(b) Determine the cubic spline you chose in part (a) and verify that it is a single cubic curve that interpolates the five points.

17.4 Repeat the calculations in the example in this chapter using a natural spline to interpolate the five data points.

17.5 Repeat the calculations in the example in this chapter using a cubic runout spline to interpolate the five data points.

17.6 (The Periodic Spline) If it is known or if it is desired that the n points (x_1, y_1), (x_2, y_2), ..., (x_n, y_n) to be interpolated lie on a single cycle of a periodic curve with period $x_n - x_1$, then an interpolating cubic spline $S(x)$ must satisfy

$$S(x_1) = S(x_n)$$
$$S'(x_1) = S'(x_n)$$
$$S''(x_1) = S''(x_n) .$$

(a) Show that these three periodicity conditions require that

$$y_1 = y_n$$
$$M_1 = M_n$$
$$4M_1 + M_2 + M_{n-1} = 6(y_{n-1} - 2y_1 + y_2)/h^2 .$$

(b) Using the three equations in part (a) and Eqs. (17.15), construct an $(n-1) \times (n-1)$ linear system for $M_1, M_2, \ldots,$ M_{n-1} in matrix form.

17.7 (The Clamped Spline) Suppose, in addition to the n points to be interpolated, we are given specific values y_1' and y_n' for the slopes $S'(x_1)$, and $S'(x_n)$ of the interpolating cubic spline at the endpoints x_1 and x_n.

(a) Show that

$$2M_1 + M_2 = 6(y_2 - y_1 - hy_1')/h^2$$
$$2M_n + M_{n-1} = 6(y_{n-1} - y_n + hy_n')/h^2 .$$

(b) Using the equations in part (a) and Eqs. (17.15), construct an $n \times n$ linear system for $M_1, M_2, \ldots, M_n$ in matrix form.

Remark: The clamped spline described in this exercise is the most accurate type of spline for interpolation work if the slopes at the endpoints are known or can be estimated.

Cryptography

A method of encoding and decoding messages using matrix transformations and modular arithmetic is discussed, and it is shown how Gaussian elimination can sometimes be used to break an opponent's code.

PREREQUISITES: Matrices
Gaussian Elimination
Linear Independence
Linear Transformations

INTRODUCTION

The study of encoding and decoding secret messages is called *cryptography*. Although secret codes date to the earliest days of written communication, there has been a recent surge of interest in the subject because of the need to maintain privacy for information transmitted over public communications lines.

In the language of cryptography, codes are called *ciphers*, uncoded messages are called *plaintext*, and coded messages are called *ciphertext*. The process of converting from plaintext to ciphertext is called *enciphering*, and the reverse process of converting from ciphertext to plaintext is called *deciphering*.

The simplest ciphers, called *substitution ciphers*, are those that replace each letter of the alphabet by a different letter. For example, in the substitution cipher

Plain   *A B C D E F G H I J K L M N O P Q R S T U V W X Y Z*

Cipher   *D E F G H I J K L M N O P Q R S T U V W X Y Z A B C*

the plaintext leter *A* is replaced by *D*, the plaintext letter *B* by *E*, and so forth. With this cipher the plaintext message

ROME WAS NOT BUILT IN A DAY

becomes

URPH ZDV·QRW EXLOW LQ D GDB

HILL CIPHERS

A disadvantage of substitution ciphers is that they preserve the frequencies of individual letters, making it relatively easy to break the code by statistical methods. One way to overcome this problem is to divide the plaintext into groups of letters and encipher the plaintext group by group, rather than one letter at a time. A system of cryptography in which the plaintext is divided into sets of n letters, each of which is replaced by a set of n cipher letters, is called a *polygraphic system*. In this section we will study a class of polygraphic systems based on matrix transformations. The ciphers that we will discuss are called Hill ciphers after Lester S. Hill who introduced them in two papers: "Cryptography in an Algebraic Alphabet," *American Mathematical Monthly, 36*, pp. 306-312, June-July 1929 and "Concerning Certain Linear Transformation Apparatus of Cryptography," *American Mathematical Monthly, 38*, pp. 135-154, March 1931.

In the discussion to follow we will assume that each plaintext and ciphertext letter except Z is assigned the numerical value that specifies its position in the standard alphabet (Table 18.1). For reasons that will become clear later, Z is assigned a value of zero.

A	*B*	*C*	*D*	*E*	*F*	*G*	*H*	*I*	*J*	*K*	*L*	*M*	*N*	*O*	*P*	*Q*	*R*	*S*	*T*	*U*	*V*	*W*	*X*	*Y*	*Z*
1	2	3	4	5	6	7	8	9	10	11	12	13	14	15	16	17	18	19	20	21	22	23	24	25	0

Table 18.1

In the simplest Hill ciphers, successive *pairs* of plaintext are transformed into ciphertext by the following procedure:

Step 1. *Choose a 2 × 2 matrix with integer entries*

$$A = \begin{bmatrix} a_{11} & a_{12} \\ a_{21} & a_{22} \end{bmatrix}$$

to perform the encoding. Certain additional conditions on A will be imposed later.

Step 2. *Group successive plaintext letters into pairs, adding an arbitrary "dummy" letter to fill out the last pair if the plaintext has an odd number of letters, and replace each plaintext letter by its numerical value.*

Step 3. *Successively convert each plaintext pair $p_1 p_2$ into a column vector*

$$p = \begin{bmatrix} p_1 \\ p_2 \end{bmatrix}$$

and form the product Ap. We will call p a plaintext vector and Ap the corresponding ciphertext vector.

Step 4. *Convert each ciphertext vector into its alphabet equivalent.*

EXAMPLE 18.1 Use the matrix

$$\begin{bmatrix} 1 & 2 \\ 0 & 3 \end{bmatrix}$$

to obtain the Hill cipher for the plaintext message

I AM HIDING

SOLUTION If we group the plaintext into pairs and add the dummy letter *G* to fill out the last pair, we obtain

IA   MH   ID   IN   GG

or equivalently from Table 18.1:

$$9\ 1 \quad 13\ 8 \quad 9\ 4 \quad 9\ 14 \quad 7\ 7$$

To encipher the pair *IA*, we form the matrix product

$$\begin{bmatrix} 1 & 2 \\ 0 & 3 \end{bmatrix} \begin{bmatrix} 9 \\ 1 \end{bmatrix} = \begin{bmatrix} 11 \\ 3 \end{bmatrix}$$

which, from Table 18.1, yields the ciphertext *KC*.
To encipher the pair *MH* we form the product

$$\begin{bmatrix} 1 & 2 \\ 0 & 3 \end{bmatrix} \begin{bmatrix} 13 \\ 8 \end{bmatrix} = \begin{bmatrix} 29 \\ 24 \end{bmatrix}. \tag{18.1}$$

However, there is a problem here because the number 29 has no
alphabet equivalent (Table 18.1). To resolve this problem we
shall make the following agreement:

*Whenever an integer greater than 25 occurs it will be
replaced by the remainder that results when this integer is divided by 26.*

Since the remainder after division by 26 is one of the integers
0, 1, 2, ... , 25, this procedure will always yield an integer
with an alphabet equivalent.
Thus, in (18.1) we replace 29 by 3, which is the remainder
after dividing 29 by 26. It now follows from Table 18.1 that the
ciphertext for the pair *MH* is *CX*.
The computations for the remaining ciphertext vectors are

$$\begin{bmatrix} 1 & 2 \\ 0 & 3 \end{bmatrix} \begin{bmatrix} 9 \\ 4 \end{bmatrix} = \begin{bmatrix} 17 \\ 12 \end{bmatrix}$$

$$\begin{bmatrix} 1 & 2 \\ 0 & 3 \end{bmatrix} \begin{bmatrix} 9 \\ 14 \end{bmatrix} = \begin{bmatrix} 37 \\ 42 \end{bmatrix} \quad \text{or} \quad \begin{bmatrix} 11 \\ 16 \end{bmatrix}$$

$$\begin{bmatrix} 1 & 2 \\ 0 & 3 \end{bmatrix} \begin{bmatrix} 7 \\ 7 \end{bmatrix} = \begin{bmatrix} 21 \\ 21 \end{bmatrix}.$$

These correspond to the ciphertext pairs *QL*, *KP*, and *UU*, respectively. In summary, the **entire** ciphertext message is

$$KC \quad CX \quad QL \quad KP \quad UU$$

which would usually be transmitted as a single string without spaces:

$$KCCXQLKPUU$$

◄◆◆◆►

   Because the plaintext was grouped in pairs and enciphered by a 2 × 2 matrix, the Hill cipher in Example 18.1 is referred to as a *Hill 2-cipher*. It is obviously also possible to group the plaintext in triples and encipher by a 3 × 3 matrix with integer entries; this is called a *Hill 3-cipher*. In general, for a *Hill n-cipher*, plaintext is grouped into sets of *n* letters and enciphered by an *n* × *n* matrix with integer entries.

MODULAR ARITHMETIC

   In Example 18.1 integers greater than 25 were replaced by their remainders after division by 26. This technique of working with remainders is at the core of a body of mathematics called *modular arithmetic*. Because of its importance in cryptography, we will digress for a moment to touch on some of the main ideas in this area.

   In modular arithmetic, one is given a positive integer *m*, called the *modulus*, and any two integers whose difference is an integer multiple of the modulus are regarded to be "equal" or "equivalent" with respect to the modulus. More precisely, we make the following definition:

DEFINITION 18.1 *If m is a positive integer and a and b are any integers, then we **say** that a is **equivalent** to b modulo m, written*

$$a = b \ (mod \ m)$$

if a − b is an integer multiple of m.

EXAMPLE 18.2

$$7 = 2 \quad (\text{mod } 5)$$
$$19 = 3 \quad (\text{mod } 2)$$
$$-1 = 25 \quad (\text{mod } 26)$$
$$12 = 0 \quad (\text{mod } 4)$$

▬◆◆◆

For any modulus m it can be proved that every integer a is equivalent, modulo m, to exactly one of the integers

$$0, 1, 2, \ldots, m-1.$$

We shall call this integer the *residue* of a modulo m, and we shall write

$$Z_m = \{0, 1, 2, \ldots, m-1\}$$

to denote the set of residues modulo m.

If a is a *nonnegative* integer then its residue modulo m is simply the remainder that results when a is divided by m. For an arbitrary integer a, the residue can be found using the following theorem:

THEOREM 18.1 *For any integer a and modulus m let*

$$R = remainder \ of \ \frac{|a|}{m} \ .$$

Then the residue r of a modulo m is given by

$$r = \begin{cases} R & if \ a \geq 0 \\ m - R & if \ a < 0 \ and \ R \neq 0 \\ 0 & if \ a < 0 \ and \ R = 0. \end{cases}$$

EXAMPLE 18.3

Find the residue modulo 26 of

(a) 87   (b) -38   (c) -26 .

SOLUTION (A)

Dividing $|87| = 87$ by 26 yields a remainder of $R = 9$, so $r = 9$. Thus,

$$87 = 9 \ (\text{mod } 26).$$

SOLUTION (B)

Dividing $|-38| = 38$ by 26 yields a remainder of $R = 12$, so $r = 26 - 12 = 14$. Thus,

$$-38 = 14 \ (\text{mod } 26).$$

SOLUTION (C)

Dividing $|-26| = 26$ by 26 yields a remainder of $R = 0$. Thus,

$$-26 = 0 \ (\text{mod } 26).$$

◆━◆━◆

In ordinary arithmetic every nonzero number a has a *reciprocal* or *multiplicative inverse*, devoted by a^{-1}, such that

$$aa^{-1} = a^{-1}a = 1.$$

In modular arithmetic we have the following corresponding concept:

> **DEFINITION 18.2** *If a is a number in Z_m, then a number a^{-1} in Z_m is called a* **reciprocal** *or* **multiplicative inverse** *of a modulo m if $aa^{-1} = a^{-1}a = 1$ (mod m).*

It can be proved that if a and m have no common prime factors, then a has a unique reciprocal modulo m, and conversely if a and m have a common prime factor then a has no reciprocal modulo m.

EXAMPLE 18.4

The number 3 has a reciprocal modulo 26 because 3 and 26 have no common prime factors. This reciprocal can be obtained by finding the number x in Z_{26} that satisfies the modular equation

$$3x = 1 \ (\text{mod } 26).$$

Although there are general methods for solving such modular equations, it would take us too far afield to study them. However, because 26 is relatively small, this equation can be solved by trying the possible solutions, 0 to 25, one at a time. With this approach we find that $x = 9$ is the solution since

$$3 \cdot 9 = 27 = 1 \pmod{26}.$$

Thus,

$$3^{-1} = 9 \pmod{26}.$$

EXAMPLE 18.5 The number 4 has no reciprocal modulo 26 because 4 and 26 have 2 as a common prime factor (see Exercise 18.8).

◆━━◆━━◆

For future reference, we provide the following table of reciprocals modulo 26:

a	1	3	5	7	9	11	15	17	19	21	23	25
a^{-1}	1	9	21	15	3	19	7	23	11	5	17	25

Table 18.2 Reciprocals Modulo 26

DECIPHERING

Every useful cipher must have a procedure for decipherment. In the case of a Hill cipher, decipherment uses the inverse (mod 26) of the enciphering matrix. To be precise, if m is a positive integer, then a square matrix A with entries in Z_m is said to be *invertible modulo m* if there is a matrix B with entries in Z_m such that

$$AB = BA = I \pmod{m}.$$

Suppose now that

$$A = \begin{bmatrix} a_{11} & a_{12} \\ a_{21} & a_{22} \end{bmatrix}$$

is invertible modulo 26 and this matrix is used in a Hill 2-cipher. If

$$\mathbf{p} = \begin{bmatrix} p_1 \\ p_2 \end{bmatrix}$$

is a plaintext vector, then

$$c = A p$$

is the corresponding ciphertext vector and

$$p = A^{-1} c.$$

Thus, each plaintext vector can be recovered from the corresponding ciphertext vector by multiplying it on the left by A^{-1} (mod m).

In cryptography it is important to know which matrices are invertible modulo 26 and how to obtain their inverses. We shall now investigate these questions.

In ordinary arithmetic, a square matrix A is invertible if and only if $\det(A) \neq 0$; or equivalently, if and only if $\det(A)$ has a reciprocal. The following theorem is the analog of this result in modular arithmetic.

THEOREM 18.2 *A square matrix A with entries in Z_m is invertible modulo m if and only if the residue of $\det(A)$ modulo m has a reciprocal modulo m.*

Since the residue of $\det(A)$ modulo m will have a reciprocal modulo m if and only if this residue and m have no common prime factors, we have the following corollary:

COROLLARY 18.1 *A square matrix A with entries in Z_m is invertible modulo m if and only if m and the residue of $\det(A)$ modulo m have no common prime factors.*

Since the only prime factors of $m = 26$ are 2 and 13, we have the following corollary that is useful in cryptography.

COROLLARY 18.2 *A square matrix A with entries in Z_{26} is invertible modulo 26 if and only if the residue of $\det(A)$ modulo 26 is not divisible by 2 or 13.*

We leave it for the reader to verify that if

$$A = \begin{bmatrix} a & b \\ c & d \end{bmatrix}$$

has entries in Z_{26} and the residue of $\det(A) = ad - bc$ modulo 26 is not divisible by 2 or 13, then the inverse of A (mod 26) is given by

$$A^{-1} = (ad - bc)^{-1} \begin{bmatrix} d & -b \\ -c & a \end{bmatrix} \pmod{26} \qquad (18.2)$$

where $(ad - bc)^{-1}$ is the multiplicative inverse of the residue of $ad - bc$ (mod 26).

EXAMPLE 18.6 Find the inverse of

$$A = \begin{bmatrix} 5 & 6 \\ 2 & 3 \end{bmatrix}$$

modulo 26.

SOLUTION

$$\det(A) = ad - bc = 5 \cdot 3 - 6 \cdot 2 = 3$$

so from Table 18.2

$$(ad - bc)^{-1} = 3^{-1} = 9 \pmod{26}.$$

Thus, from (18.2)

$$A^{-1} = 9 \begin{bmatrix} 3 & -6 \\ -2 & 5 \end{bmatrix} = \begin{bmatrix} 27 & -54 \\ -18 & 45 \end{bmatrix} = \begin{bmatrix} 1 & 24 \\ 8 & 19 \end{bmatrix} \pmod{26}.$$

As a check,

$$AA^{-1} = \begin{bmatrix} 5 & 6 \\ 2 & 3 \end{bmatrix} \begin{bmatrix} 1 & 24 \\ 8 & 19 \end{bmatrix} = \begin{bmatrix} 53 & 234 \\ 26 & 105 \end{bmatrix} = \begin{bmatrix} 1 & 0 \\ 0 & 1 \end{bmatrix} \quad (\text{mod } 26).$$

Similarly, $A^{-1}A = I$.

EXAMPLE 18.7

Decode the following Hill 2-cipher which was enciphered by the matrix in Example 18.6:

GTNKGKDUSK

SOLUTION

From Table 18.1 the numerical equivalent of this ciphertext is

$$7 \ 20 \quad 14 \ 11 \quad 7 \ 11 \quad 4 \ 21 \quad 19 \ 11$$

To obtain the plaintext pairs, we multiply each ciphertext vector by the inverse of A (obtained in Example 18.5):

$$\begin{bmatrix} 1 & 24 \\ 8 & 19 \end{bmatrix} \begin{bmatrix} 7 \\ 20 \end{bmatrix} = \begin{bmatrix} 487 \\ 436 \end{bmatrix} = \begin{bmatrix} 19 \\ 20 \end{bmatrix} \quad (\text{mod } 26)$$

$$\begin{bmatrix} 1 & 24 \\ 8 & 19 \end{bmatrix} \begin{bmatrix} 14 \\ 11 \end{bmatrix} = \begin{bmatrix} 278 \\ 321 \end{bmatrix} = \begin{bmatrix} 18 \\ 9 \end{bmatrix} \quad (\text{mod } 26)$$

$$\begin{bmatrix} 1 & 24 \\ 8 & 19 \end{bmatrix} \begin{bmatrix} 7 \\ 11 \end{bmatrix} = \begin{bmatrix} 271 \\ 265 \end{bmatrix} = \begin{bmatrix} 11 \\ 5 \end{bmatrix} \quad (\text{mod } 26)$$

$$\begin{bmatrix} 1 & 24 \\ 8 & 19 \end{bmatrix} \begin{bmatrix} 4 \\ 21 \end{bmatrix} = \begin{bmatrix} 508 \\ 431 \end{bmatrix} = \begin{bmatrix} 14 \\ 15 \end{bmatrix} \quad (\text{mod } 26)$$

$$\begin{bmatrix} 1 & 24 \\ 8 & 19 \end{bmatrix} \begin{bmatrix} 19 \\ 11 \end{bmatrix} = \begin{bmatrix} 283 \\ 361 \end{bmatrix} = \begin{bmatrix} 23 \\ 23 \end{bmatrix} \quad (\text{mod } 26).$$

From Table 18.1 the alphabet equivalents of these vectors are

$$ST \quad RI \quad KE \quad NO \quad WW$$

which yields the message

$$STRIKE \ NOW$$

BREAKING A HILL CIPHER

Since the purpose of enciphering messages and information is to prevent the "opponents" from learning their contents, cryptographers are concerned with the *security* of their ciphers; that is, how readily they can be broken (deciphered by their opponents). We will conclude this chapter by discussing one technique for breaking Hill ciphers.

Suppose that you are able to obtain some corresponding plaintext and ciphertext from an opponent's message. For example, on examining some intercepted ciphertext you may be able to deduce that the message is a letter that begins *DEAR SIR*. We will show that with a small amount of such data it may be possible to determine the deciphering matrix of a Hill code and consequently obtain access to the rest of the message.

It is a basic result in linear algebra that a linear transformation is completely determined by its values at a basis. This principle suggests that if we have a Hill n-cipher, and if

$$p_1, \ p_2, \ldots, \ p_n$$

are linearly independent plaintext vectors whose corresponding ciphertext vectors

$$Ap_1, \ Ap_2, \ldots, \ Ap_n$$

are known, then there is enough information available to determine the matrix A and hence A^{-1} (mod m).

The following theorem, whose proof is discussed in the exercises, will provide a way to do this.

THEOREM 18.3 *Let* p_1, p_2, ..., p_n *be linearly independent plaintext vectors, and let* c_1, c_2, ..., c_n *be the corresponding ciphertext vectors in a Hill n-cipher. If*

$$
P = \begin{bmatrix} p_1^t \\ p_2^t \\ \vdots \\ p_n^t \end{bmatrix}
$$

is the $n \times n$ matrix with row vectors p_1^t, p_2^t, ..., p_n^t and if

$$
C = \begin{bmatrix} c_1^t \\ c_2^t \\ \vdots \\ c_n^t \end{bmatrix}
$$

is the $n \times n$ matrix with row vectors c_1^t, c_2^t, ..., c_n^t, then the sequence of elementary row operations that reduces C to I, transforms P to $(A^{-1})^t$.

This theorem tells us that to find the transpose of the deciphering matrix A^{-1}, we must find a sequence of row operations that reduces C to I and then perform this same sequence of operations on P. The following example illustrates a simple algorithm for doing this.

EXAMPLE 18.8 The following Hill 2-cipher is intercepted:

IOSBTGXESPXHOPDE

Decipher the message given that it starts with the word *DEAR*.

SOLUTION From Table 18.1, the numerical equivalent of the known plaintext is

$$\begin{array}{cc} DE & AR \\ 4\ \ 5 & 1\ \ 18 \end{array}$$

and the numerical equivalent of the corresponding ciphertext is

$$\begin{array}{cc} IO & SB \\ 9\ \ 15 & 19\ \ 2 \end{array}$$

so the corresponding plaintext and ciphertext vectors are

$$\mathbf{p}_1 = \begin{bmatrix} 4 \\ 5 \end{bmatrix} \iff \mathbf{c}_1 = \begin{bmatrix} 9 \\ 15 \end{bmatrix}$$

$$\mathbf{p}_2 = \begin{bmatrix} 1 \\ 18 \end{bmatrix} \iff \mathbf{c}_2 = \begin{bmatrix} 19 \\ 2 \end{bmatrix}.$$

We want to reduce

$$C = \begin{bmatrix} \mathbf{c}_1^t \\ \mathbf{c}_2^t \end{bmatrix} = \begin{bmatrix} 9 & 15 \\ 19 & 2 \end{bmatrix}$$

to I by elementary row operations and simultaneously apply these operations to

$$P = \begin{bmatrix} \mathbf{p}_1^t \\ \mathbf{p}_2^t \end{bmatrix} = \begin{bmatrix} 4 & 5 \\ 1 & 18 \end{bmatrix}$$

to obtain $(A^{-1})^t$ (the transpose of the deciphering matrix). This can be accomplished by adjoining P to the right of C and applying row operations to the resulting matrix $[C|P]$ until the left side is reduced to I. The final matrix will then have the form $[I|(A^{-1})^t]$. The computations can be carried out as follows:

$$\begin{bmatrix} 9 & 15 & | & 4 & 5 \\ 19 & 2 & | & 1 & 18 \end{bmatrix}$$

We formed the matrix $[C|P]$.

$$\begin{bmatrix} 1 & 45 & | & 12 & 15 \\ 19 & 2 & | & 1 & 18 \end{bmatrix}$$

We multiplied the first row by $9^{-1} = 3$.

$$\begin{bmatrix} 1 & 19 & | & 12 & 15 \\ 19 & 2 & | & 1 & 18 \end{bmatrix}$$

We replaced 45 by its residue modulo 26.

$$\begin{bmatrix} 1 & 19 & | & 12 & 15 \\ 0 & -359 & | & -227 & -267 \end{bmatrix}$$

We added -19 times the first row to the second.

$$\begin{bmatrix} 1 & 19 & | & 12 & 15 \\ 0 & 5 & | & 7 & 19 \end{bmatrix}$$

We replaced the entries in the second row by their residues modulo 26.

$$\begin{bmatrix} 1 & 19 & | & 12 & 15 \\ 0 & 1 & | & 147 & 399 \end{bmatrix}$$

We multiplied the second row by $5^{-1} = 21$.

$$\begin{bmatrix} 1 & 19 & | & 12 & 15 \\ 0 & 1 & | & 17 & 9 \end{bmatrix}$$

We replaced the entries in the second row by their residues modulo 26.

$$\begin{bmatrix} 1 & 0 & | & -311 & -156 \\ 0 & 1 & | & 17 & 9 \end{bmatrix}$$

We added -19 times the second row to the first.

$$\begin{bmatrix} 1 & 0 & | & 1 & 0 \\ 0 & 1 & | & 17 & 9 \end{bmatrix}$$

We replaced the entries in the first row by their residues modulo 26.

Thus,

$$(A^{-1})^t = \begin{bmatrix} 1 & 0 \\ 17 & 9 \end{bmatrix}$$

so the deciphering matrix is

$$A^{-1} = \begin{bmatrix} 1 & 17 \\ 0 & 9 \end{bmatrix}.$$

To decipher the message, we first group the ciphertext into pairs and find the numerical equivalent of each letter:

IO	SB	TG	XE	SP	XH	OP	DE
9 15	19 2	20 7	24 5	19 16	24 8	15 16	4 5

Next, we multiply successive ciphertext vectors on the left by A^{-1} and find the alphabet equivalents of the resulting plaintext pairs:

$$\begin{bmatrix} 1 & 17 \\ 0 & 9 \end{bmatrix}\begin{bmatrix} 9 \\ 15 \end{bmatrix} = \begin{bmatrix} 4 \\ 5 \end{bmatrix} \quad \begin{matrix} D \\ E \end{matrix}$$

$$\begin{bmatrix} 1 & 17 \\ 0 & 9 \end{bmatrix}\begin{bmatrix} 19 \\ 2 \end{bmatrix} = \begin{bmatrix} 1 \\ 18 \end{bmatrix} \quad \begin{matrix} A \\ R \end{matrix}$$

$$\begin{bmatrix} 1 & 17 \\ 0 & 9 \end{bmatrix}\begin{bmatrix} 20 \\ 7 \end{bmatrix} = \begin{bmatrix} 9 \\ 11 \end{bmatrix} \quad \begin{matrix} I \\ K \end{matrix} \qquad (\text{mod } 26)$$

$$\begin{bmatrix} 1 & 17 \\ 0 & 9 \end{bmatrix}\begin{bmatrix} 24 \\ 5 \end{bmatrix} = \begin{bmatrix} 5 \\ 19 \end{bmatrix} \quad \begin{matrix} E \\ S \end{matrix}$$

$$\begin{bmatrix} 1 & 17 \\ 0 & 9 \end{bmatrix}\begin{bmatrix} 19 \\ 16 \end{bmatrix} = \begin{bmatrix} 5 \\ 14 \end{bmatrix} \quad \begin{matrix} E \\ N \end{matrix}$$

$$\begin{bmatrix} 1 & 17 \\ 0 & 9 \end{bmatrix}\begin{bmatrix} 24 \\ 8 \end{bmatrix} = \begin{bmatrix} 4 \\ 20 \end{bmatrix} \quad \begin{matrix} D \\ T \end{matrix}$$

$$\begin{bmatrix} 1 & 17 \\ 0 & 9 \end{bmatrix}\begin{bmatrix} 15 \\ 16 \end{bmatrix} = \begin{bmatrix} 1 \\ 14 \end{bmatrix} \quad \begin{matrix} A \\ N \end{matrix}$$

$$\begin{bmatrix} 1 & 17 \\ 0 & 9 \end{bmatrix}\begin{bmatrix} 4 \\ 5 \end{bmatrix} = \begin{bmatrix} 11 \\ 19 \end{bmatrix} \quad \begin{matrix} K \\ S \end{matrix}$$

(mod 26)

Finally, we construct the message from the plaintext pairs:

DE AR IK ES EN DT AN KS

DEAR IKE SEND TANKS

FURTHER READINGS

Readers interested in learning more about mathematical cryptography are referred to the following books, the first of which is elementary and the second more advanced:

[1] Abraham Sinkov, *Elementary Cryptanalysis, a Mathematical Approach*, Mathematical Library, Mathematical Association of America, 1966.

[2] Alan G. Konheim, *Cryptography, a Primer*, Wiley-Interscience, 1981.

EXERCISES

18.1 Obtain the Hill cipher of the message

DARK NIGHT

for each of the following enciphering matrices:

(a) $\begin{bmatrix} 1 & 3 \\ 2 & 1 \end{bmatrix}$ 　　　　　(b) $\begin{bmatrix} 4 & 3 \\ 1 & 2 \end{bmatrix}$.

18.2 In each part determine whether the matrix is invertible modulo 26. If so, find its inverse modulo 26 and check your work by verifying that $AA^{-1} = A^{-1}A = I$ (mod 26).

(a) $A = \begin{bmatrix} 9 & 1 \\ 7 & 2 \end{bmatrix}$ (b) $A = \begin{bmatrix} 3 & 1 \\ 5 & 3 \end{bmatrix}$ (c) $A = \begin{bmatrix} 8 & 11 \\ 1 & 9 \end{bmatrix}$

(d) $A = \begin{bmatrix} 2 & 1 \\ 1 & 7 \end{bmatrix}$ (e) $A = \begin{bmatrix} 3 & 1 \\ 6 & 2 \end{bmatrix}$ (f) $A = \begin{bmatrix} 1 & 8 \\ 1 & 3 \end{bmatrix}$.

18.3 Decode the message

$$SAKNOXAOJX$$

given that it is a Hill cipher with enciphering matrix

$$\begin{bmatrix} 4 & 1 \\ 3 & 2 \end{bmatrix} .$$

18.4 A Hill 2-cipher is intercepted which starts with the pairs

$$SL \quad HK$$

Find the deciphering and enciphering matrices given that the plaintext is known to start with the word *ARMY*.

18.5 Decode the following Hill 2-cipher if the last four plaintext letters are known to be *ATOM*.

$$LNGIHGYBVRENJYQO$$

18.6 Decode the following Hill 3-cipher if the first nine plaintext letters are *IHAVECOME* :

$$HPAFQGGDUGDDHPGODYNOR$$

18.7 If, in addition to the standard alphabet, a period, comma, and question mark are allowed, then 29 plaintext and ciphertext symbols are available and all matrix arithmetic would be done modulo 29. Under what conditions would a matrix with entries in Z_{29} be invertible modulo 29?

18.8 Show that the modular equation $4x = 1$ (mod 26) has no solution in Z_{26} by successively substituting the values $x = 0, 1, 2, \ldots, 25$.

18.9 (a) Let P and C be the matrices in Theorem 18.3. Show that $P = C(A^{-1})^t$.

(b) To prove Theorem 18.3, let $E_1, E_2, \ldots, E_n$ be the elementary matrices that correspond to the row operations which reduce C to I, so

$$E_n \cdots E_2 E_1 C = I.$$

Show that

$$E_n \cdots E_2 E_1 P = (A^{-1})^t$$

from which it follows that the same sequence of row operations that reduces C to I converts P to $(A^{-1})^t$.

18.10 (a) If A is the enciphering matrix of a Hill n-cipher, show that

$$A^{-1} = (C^{-1}P)^t \qquad \text{(mod 26)}$$

where C and P are the matrices defined in Theorem 18.3.

(b) Instead of using Theorem 18.3 as in the text, find the deciphering matrix A^{-1} of Example 18.8 by using the result in part (a) and Eq. (18.2) to compute C^{-1}.

Remark: Although this method is practical for Hill 2-ciphers, Theorem 18.3 is more efficient for Hill n-ciphers with $n > 2$.

Computed Tomography

Constructing a cross-sectional view of a human body by analyzing x-ray scans leads to an inconsistent overdetermined linear system for the digital representation of the view. An iteration technique is presented that provides an "approximate solution" of the linear system.

PREREQUISITES: Linear systems

Euclidean space R^n

INTRODUCTION

The basic problem of computed tomography is the construction of an image of a cross-section of the human body using data collected from many individual beams of x-rays that are passed through the cross-section. These data are processed by a computer and the computed cross-section is displayed on a video monitor. Figure 19.1 is a diagram of General Electric's CT/T system showing a patient prepared to have a cross-section of his head scanned by x-ray beams. Such a system is also known as a *CAT-scanner*, for Computer-Aided-Tomography-scanner. Figure 19.2 displays a typical cross-section of a human head produced by the system.

Figure 19.1

Figure 19.2

The first commercial system of computed tomography for medical use was developed in 1971 by G. N. Hounsfield of EMI, Ltd. in England. In 1979, Hounsfield and A. M. Cormack were awarded the Nöbel prize for their pioneering work in the field. As we shall see in this chapter, the construction of a cross-section or tomograph (*Greek* tomos - τομος: a cut or slice) requires the solution of a large linear system of equations. Certain algorithms, called Algebraic Reconstruction Techniques (ART), can be used to solve these linear systems whose solutions yield the cross-sections in digital form.

Scanning Modes

Unlike conventional x-ray pictures which are formed by x-rays that are projected *perpendicular* to the plane of the picture, tomographs are constructed from thousands of individual, hairline-thin x-ray beams that *lie in the plane* of the cross-section. After they pass through the cross-section, the intensity of the x-ray beams are measured by an x-ray detector and these measurements are relayed to a computer where they are processed. Figures 19.3 and 19.4 illustrate two possible modes of scanning the cross-section: the *parallel mode* and the *fan-beam mode*. In the parallel mode, a single x-ray source and x-ray detector pair are translated across the field of view containing the cross-section and many measurements of the parallel beams are recorded. Then the source and

Figure 19.3

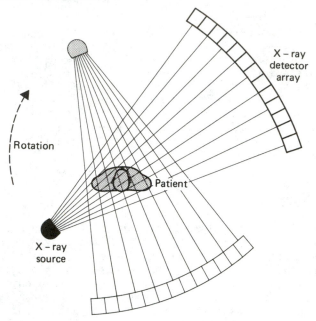

X – ray
detector
array

Rotation

Patient

X – ray
source

Figure 19.4

detector pair are rotated through a small angle and another set
of measurements is taken. This is repeated until the desired
number of beam measurements is completed. For example, in the
original 1971 machine, 160 parallel measurements were taken
through 180 angles spaced 1° apart: a total of $160 \times 180 = 28,800$
beam measurements. Each such scan took approximately $5\frac{1}{2}$ minutes.

In the fan-beam mode of scanning, a single x-ray tube generates a fan of collimated beams whose intensities are measured
simultaneously by an array of detectors on the other side of the
field of view. The x-ray tube and detector array are rotated
through many angles and a set of measurements is taken at each
angle until the scan is completed. In the General Electric CT/T
system, which uses the fan-beam mode, each scan takes 4.6 seconds
and a total of 184,320 beam measurements are recorded.

DERIVATION OF EQUATIONS

To see how the cross-section is reconstructed from the many
individual beam measurements, refer to Figure 19.5. Here the
field of view in which the cross-section is situated has been
divided into many square *pixels* (picture elements) numbered 1
through N as indicated. It is our desire to determine the x-ray
density of each pixel. In the EMI system, 6,400 pixels were used,
arranged in a square 80×80 array. The G.E. CT/T system uses

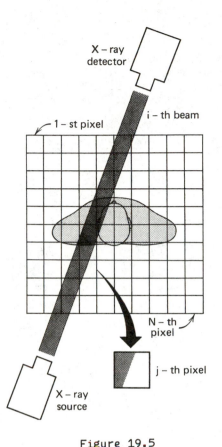

Figure 19.5

102,400 pixels in a 320×320 array, each pixel being about 1 mm on a side. After the densities of the pixels are determined by the method we shall describe, they are reproduced on a video monitor with each pixel shaded a level of gray proportional to its x-ray density. Since different tissues within the human body have different x-ray densities, the video display clearly distinguishes the various tissues and organs within the cross-section.

Figure 19.6 shows a single pixel with an x-ray beam of roughly the same width as the pixel passing squarely through it. The photons constituting the x-ray beam are absorbed by the tissue within the pixel at a rate proportional to the x-ray density of the tissue. Quantitatively, the x-ray density of the jth pixel is denoted by x_j and is defined by

Figure 19.6

$$x_j = ln\left(\frac{number\ of\ photons\ entering\ the\ jth\ pixel}{number\ of\ photons\ leaving\ the\ jth\ pixel}\right)$$

where "ln" denotes the natural logarithmic function. Using the fact that $ln(a/b) = -ln(b/a)$, we also have

$$x_j = -ln\left(\begin{array}{l}fraction\ of\ photons\ that\ pass\ through\ the\ full\\ width\ of\ the\ jth\ pixel\ without\ being\ absorbed\end{array}\right).$$

If the x-ray beam passes through an entire row of pixels
(Fig. 19.7) then the number of photons leaving one pixel is equal
to the number of photons entering the next pixel in the row. If
the pixels are numbered 1, 2, ... , n, then the additive property
of the logarithmic function gives

$$x_1 + x_2 + \cdots + x_n = ln \left(\frac{number\ of\ photons\ entering\ the\ 1st\ pixel}{number\ of\ photons\ leaving\ the\ nth\ pixel} \right)$$

$$= -ln \left(\begin{array}{l} fraction\ of\ photons\ that\ pass \\ through\ the\ row\ of\ n\ pixels \\ without\ being\ absorbed \end{array} \right). \qquad (19.1)$$

Photons entering
1—st pixel

Photons leaving
n—th pixel

Figure 19.7

Thus, to determine the total x-ray density of a row of pixels, we
simply sum the individual pixel densities.

Next, consider the x-ray beam in Figure 19.5. By the *beam
density* of the ith beam of a scan, denoted by b_i, we mean

$$b_i = ln \left(\begin{array}{l} number\ of\ photons\ of\ the\ ith\ beam\ entering\ the\ detec\text{-} \\ tor\ without\ the\ cross\text{-}section\ in\ the\ field\ of\ view \\ \overline{number\ of\ photons\ of\ the\ ith\ beam\ entering\ the\ detec\text{-}} \\ tor\ with\ the\ cross\text{-}section\ in\ the\ field\ of\ view \end{array} \right)$$

$$= -ln \left(\begin{array}{l} fraction\ of\ photons\ of\ the\ ith\ beam\ that \\ pass\ through\ the\ cross\text{-}section\ without \\ being\ absorbed \end{array} \right). \qquad (19.2)$$

The numerator in the first expression for b_i is obtained by performing a calibration scan without the cross-section in the field
of view. The resulting detector measurements are stored within
the computer's memory. Then a clinical scan is performed with
the cross-section in the field of view. The clinical measurements are divided into the calibration measurements, the b_i's
of all the beams constituting the scan are computed, and the
values are stored for further processing.

For each beam that passes squarely through a row of pixels,
we must have

$$\begin{pmatrix} \text{fraction of photons of the} \\ \text{beam that pass through the} \\ \text{row of pixels without being} \\ \text{absorbed} \end{pmatrix} = \begin{pmatrix} \text{fraction of photons of the} \\ \text{beam that pass through the} \\ \text{cross-section without being} \\ \text{absorbed} \end{pmatrix}$$

Thus, if the ith beam passes squarely through a row of n pixels, then it follows from Eqs. (19.1) and (19.2) that

$$x_1 + x_2 + \cdots + x_n = b_i .$$

In this equation, b_i is known from the clinical and calibration measurements and x_1, x_2, $\ldots$, x_n are unknown pixel densities that must be determined.

More generally, if the ith beam passes squarely through a row (or column) of pixels with numbers j_1, j_2, $\ldots$, j_i, then we have

$$x_{j_1} + x_{j_2} + \cdots + x_{j_i} = b_i$$

If we set

$$a_{ij} = \begin{cases} 1 & \text{if } j = j_1, \ j_2, \ \ldots, \ j_i \\ 0 & \text{otherwise,} \end{cases}$$

then we may write this equation as

$$a_{i1}x_1 + a_{i2}x_2 + \cdots + a_{iN}x_N = b_i . \tag{19.3}$$

We shall refer to Eq. (19.3) as the *ith beam equation*.

Referring to Figure 19.5, however, we see that the beams of a scan do not necessarily pass through a row or column of pixels squarely. Instead, a typical beam passes diagonally through each pixel in its path. There are many ways to take this into account. On the next page we outline three methods of defining the quantities a_{ij} that appear in the ith beam equation (19.3), each of which reduces to our previous definition when the beam passes squarely through a row or column of pixels. Reading down the page, each method is more exact than its predecessor, but with successively more computational difficulty.

CENTER-OF-PIXEL METHOD

$$a_{ij} = \begin{cases} 1 & \textit{if the ith beam passes} \\ & \textit{through the center of} \\ & \textit{the jth pixel} \\ 0 & \textit{otherwise} \end{cases}$$

*i*th beam

*j*th pixel

CENTER LINE METHOD

$$a_{ij} = \left(\frac{\begin{array}{l}\textit{length of the center line}\\\textit{of the ith beam that lies}\\\textit{in the jth pixel}\end{array}}{\textit{width of the jth pixel}} \right)$$

length of
center line

width of
pixel

AREA METHOD

$$a_{ij} = \left(\frac{\textit{area of the ith beam that lies in the jth pixel}}{\begin{array}{l}\textit{area of the ith beam that would lie in the jth pixel}\\\textit{if the ith beam were to cross the pixel squarely}\end{array}} \right)$$

Area in numerator
of a_{ij}

Area in denominator
of a_{ij}

Using any one of the three methods to define the a_{ij}'s in the ith beam equation, the set of M beam equations in a complete scan can be written as

$$a_{11}x_1 + a_{12}x_2 + \cdots + a_{1N}x_N = b_1$$
$$a_{21}x_1 + a_{22}x_2 + \cdots + a_{2N}x_N = b_2$$
$$\vdots \qquad\qquad \vdots$$
$$a_{M1}x_1 + a_{M2}x_2 + \cdots + a_{MN}x_N = b_M .$$

(19.4)

In this way we have a linear system of M equations (the M beam equations) in N unknowns (the N pixel densities).

Depending on the number of beams and pixels used, we may have $M > N$, $M = N$, or $M < N$. We shall only consider the case $M > N$, the so-called *overdetermined case* in which there are more beams in the scan than pixels in the field of view. Because of inherent modeling and experimental errors in the problem, we should not expect our linear system to have an exact mathematical solution for the pixel densities. In the next section we shall attempt to find an "approximate" solution to this linear system.

ALGEBRAIC RECONSTRUCTION TECHNIQUES

There have been many mathematical algorithms devised to treat the overdetermined linear system (19.4). The one we shall describe belongs to the class of so-called *Algebraic Reconstruction Techniques* (ART). This method, which can be traced to an iterative technique originally introduced by S. Kaczmarz in 1937, was the one used in the first commercial machine. To introduce this technique, consider the following system of three equations in two unknowns:

$$L_1: \quad x_1 + x_2 = 2$$
$$L_2: \quad x_1 - 2x_2 = -2$$
$$L_3: \quad 3x_1 - x_2 = 3 .$$

(19.5)

The lines L_1, L_2, L_3 determined by these three equations are plotted in the x_1x_2 - plane (Fig. 19.8(a)). As shown in the figure,

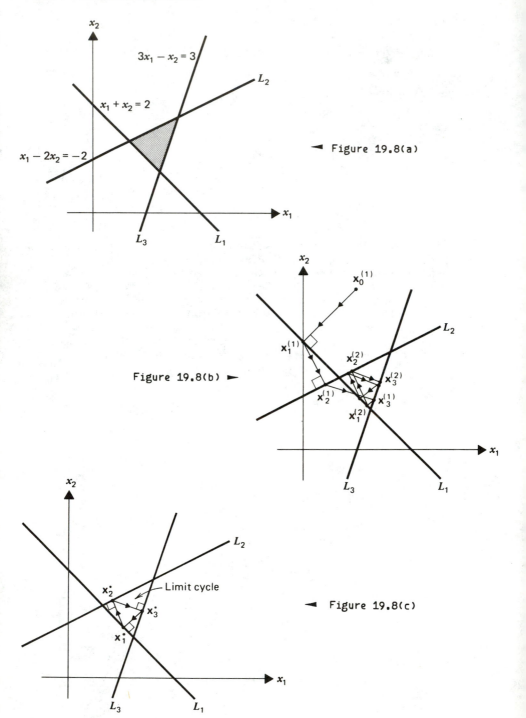

Figure 19.8(a)

Figure 19.8(b) ►

◄ Figure 19.8(c)

the three lines do not have a common intersection and so the three equations do not have an exact solution. However, the points (x_1, x_2) on the shaded triangle formed by the three lines are all situated "near" these three lines and can be thought of as constituting "approximate" solutions to our system. The following iterative procedure describes a geometric construction for generating points on the boundary of that triangular region (Fig. 19.8(b)):

ALGORITHM 19.1

STEP 0 *Choose an arbitrary starting point $x_0^{(1)}$ in the $x_1 x_2$-plane.*

STEP 1 *Project $x_0^{(1)}$ orthogonally onto the first line L_1 and call the projection $x_1^{(1)}$. The superscript (1) indicates that this is the first of several cycles through the steps.*

STEP 2 *Project $x_1^{(1)}$ orthogonally onto the second line L_2 and call the projection $x_2^{(1)}$.*

STEP 3 *Project $x_2^{(1)}$ orthogonally onto the third line L_3 and call the projection $x_3^{(1)}$.*

STEP 4 *Redefine $x_0^{(1)}$ by $x_0^{(1)} = x_3^{(1)}$ and cycle through Steps 1 through 3 again. In the second cycle, label the projected points $x_1^{(2)}$, $x_2^{(2)}$, $x_3^{(2)}$, in the third cycle label the projected points $x_1^{(3)}$, $x_2^{(3)}$, $x_3^{(3)}$, and so forth.*

This algorithm generates three sequences of points,

$$L_1: \quad x_1^{(1)}, \; x_1^{(2)}, \; x_1^{(3)}, \; \ldots$$

$$L_2: \quad x_2^{(1)}, \; x_2^{(2)}, \; x_2^{(3)}, \; \ldots$$

$$L_3: \quad x_3^{(1)}, \; x_3^{(2)}, \; x_3^{(3)}, \; \ldots$$

that lie on the three lines L_1, L_2, and L_3, respectively. It can be shown that as long as the three lines are not all parallel, then the first sequence converges to a point x_1^* on L_1, the second sequence converges to a point x_2^* on L_2, and the third sequence converges to a point x_3^* on L_3 (Fig. 19.8(c)). These three limit points form what is called the *limit cycle* of the iterative process. It can be shown that the limit cycle is independent of the starting point $x_0^{(1)}$.

We next discuss the specific formulas needed to effect the orthogonal projections in Algorithm 19.1. First, since the equation of a line in x_1x_2-space is

$$a_1 x_1 + a_2 x_2 = b,$$

we may express it in vector form as

$$a^t x = b$$

where

$$a = \begin{bmatrix} a_1 \\ a_2 \end{bmatrix} \quad \text{and} \quad x = \begin{bmatrix} x_1 \\ x_2 \end{bmatrix}$$

The following theorem from Linear Algebra gives the necessary projection formula (Exercise 19.5).

THEOREM 19.1 *Let L be a line in* R^2 *with equation* $\mathbf{a}^t\mathbf{x} = b$ *and let* $\mathbf{x}^*$ *be any point in* R^2*. Then the orthogonal projection,* $\mathbf{x}_p$*, of* $\mathbf{x}^*$ *onto L is given by*

$$\mathbf{x}_p = \mathbf{x}^* + \frac{(b - \mathbf{a}^t\mathbf{x}^*)}{\mathbf{a}^T\mathbf{a}}\,\mathbf{a}.$$

Figure 19.9

EXAMPLE 19.1 We shall use Algorithm 19.1 to find an approximate solution of the linear system given in Eq. (19.5) and illustrated in Figure 19.8. If we write the equations of the three lines as

$$L_1:\ \mathbf{a}_1^t\mathbf{x} = b_1$$

$$L_2:\ \mathbf{a}_2^t\mathbf{x} = b_2$$

$$L_3:\ \mathbf{a}_3^t\mathbf{x} = b_3$$

where

$$\mathbf{x} = \begin{bmatrix} x_1 \\ x_2 \end{bmatrix}, \quad \mathbf{a}_1 = \begin{bmatrix} 1 \\ 1 \end{bmatrix}, \quad \mathbf{a}_2 = \begin{bmatrix} 1 \\ -2 \end{bmatrix}, \quad \mathbf{a}_3 = \begin{bmatrix} 3 \\ -1 \end{bmatrix},$$

$$b_1 = 2, \qquad b_2 = -2, \qquad b_3 = 3,$$

then using Theorem 19.1, the iteration scheme in Algorithm 19.1 can be expressed as

$$x_k^{(p)} = x_{k-1}^{(p)} + \frac{b_k - \mathbf{a}_k^t x_{k-1}^{(p)}}{\mathbf{a}_k^t \mathbf{a}_k}\,\mathbf{a}_k, \qquad k = 1,\ 2,\ 3$$

where $p = 1$ for the first cycle
of iterates, $p = 2$ for the second
cycle of iterates, and so forth.
After each cycle of iterates (i.e.,
after $x_3^{(p)}$ is computed) the next
cycle of iterates is begun with
$x_0^{(p+1)}$ set equal to $x_3^{(p)}$.

Table 19.1 gives the numerical results of six cycles of
iterations starting with the
initial point $x_0^{(1)} = (1, 3)$.

Using certain techniques
that are impractical for large
linear systems, the exact values
of the points of the limit cycle
in this example can be shown to
be

$$x_1^* = (12/11, \ 10/11)$$
$$= (1.09090..., \ .90909...)$$

$$x_2^* = (46/55, \ 78/55)$$
$$= (\ .83636..., \ 1.41818...)$$

$$x_3^* = (31/22, \ 27/22)$$
$$= (1.40909..., \ 1.22727...) \ .$$

It can be seen that the sixth
cycle of iterates provides an
excellent approximation to the
limit cycle. Any one of the
three iterates $x_1^{(6)}$, $x_2^{(6)}$, $x_3^{(6)}$

Table 19.1

	x_1	x_2
$x_0^{(1)}$	1.00000	3.00000
$x_1^{(1)}$	.00000	2.00000
$x_2^{(1)}$	.40000	1.20000
$x_3^{(1)}$	1.30000	.90000
$x_1^{(2)}$	1.20000	.80000
$x_2^{(2)}$	.88000	1.44000
$x_3^{(2)}$	1.42000	1.26000
$x_1^{(3)}$	1.08000	.92000
$x_2^{(3)}$	.83200	1.41600
$x_3^{(3)}$	1.40800	1.22400
$x_1^{(4)}$	1.09200	.90800
$x_2^{(4)}$	.83680	1.41840
$x_3^{(4)}$	1.40920	1.22760
$x_1^{(5)}$	1.09080	.90920
$x_2^{(5)}$	.83632	1.41816
$x_3^{(5)}$	1.40908	1.22724
$x_1^{(6)}$	1.09092	.90908
$x_2^{(6)}$	.83637	1.41818
$x_3^{(6)}$	1.40909	1.22728

can be used as an approximate solution of the linear system. (The
large discrepancies in the values of $x_1^{(6)}$, $x_2^{(6)}$, and $x_3^{(6)}$ are due
to the artificial nature of this illustrative example. In practi-
cal problems these discrepancies would be much smaller.)

To generalize Algorithm 19.1 to treat an overdetermined system of M equations in N unknowns,

$$a_{11}x_1 + a_{12}x_2 + \cdots + a_{1N}x_N = b_1$$

$$a_{21}x_1 + a_{22}x_2 + \cdots + a_{2N}x_N = b_2 \tag{19.6}$$

$$\vdots \qquad \qquad \vdots$$

$$a_{M1}x_1 \quad a_{M2}x_2 + \cdots + a_{MN}x_N = b_M,$$

we introduce column vectors $\mathbf{x}$ and $\mathbf{a}_i$ as follows:

$$\mathbf{x} = \begin{bmatrix} x_1 \\ x_2 \\ \vdots \\ x_N \end{bmatrix}, \quad \mathbf{a}_i = \begin{bmatrix} a_{i1} \\ a_{i2} \\ \vdots \\ a_{iN} \end{bmatrix}, \quad i = 1, 2, \ldots, M.$$

With these vectors, the M equations constituting our linear system (19.6) can be written in vector form as

$$\mathbf{a}_i^t \mathbf{x} = b_i, \quad i = 1, 2, \ldots, M.$$

Each of these M equations defines what is called a *hyperplane* in the N-dimensional Euclidean space R^N. In general, these M hyperplanes have no common intersection, and so we seek instead some point in R^N that is reasonably "close" to all of them. Such a point will constitute an approximate solution of the linear system and its N entries will determine approximate pixel densities with which to form the desired cross-section.

As in the 2-dimensional case, we shall introduce an iterative process that generates cycles of successive orthogonal projections onto the M hyperplanes beginning with some arbitrary initial point in R^N. Our notation for these successive iterates is

$$\mathbf{x}_k^{(p)} = \begin{pmatrix} \textit{the iterate lying on the kth hyper-} \\ \textit{plane generated during the pth cycle} \\ \textit{of iterations.} \end{pmatrix}$$

The algorithm is as follows:

ALGORITHM 19.2

STEP 0 *Choose any point in R^N and label it $x_0^{(1)}$.*

STEP 1 *For the first cycle of iterates, set $p = 1$.*

STEP 2 *For $k = 1, 2, \ldots, M$, compute*

$$x_k^{(p)} = x_{k-1}^{(p)} + \frac{(b_k - a_k^t x_{k-1}^{(p)})}{a_k^t a_k} a_k.$$

STEP 3 *Set $x_0^{(p+1)} = x_M^{(p)}$.*

STEP 4 *Increase the cycle number p by 1 and return to Step 2.*

In Step 2, the iterate $x_k^{(p)}$ is called the *orthogonal projection* of $x_{k-1}^{(p)}$ onto the hyperplane $a_k^t x = b_k$. Consequently, as in the 2-dimensional case, this algorithm determines a sequence of orthogonal projections from one hyperplane onto the next in which we cycle back to the first hyperplane after each projection onto the last hyperplane.

It can be shown that if the vectors $a_1, a_2, \ldots, a_M$ span R^N, then the iterates $x_M^{(1)}, x_M^{(2)}, x_M^{(3)}, \ldots$ lying on the Mth hyperplane will converge to a point x_M^* on that hyperplane that does not depend on the choice of the initial point $x_0^{(1)}$. In computed tomography, one of the iterates $x_M^{(p)}$ for p sufficiently large is taken as an approximate solution of the linear system for the pixel densities.

Notice that for the center-of-pixel method the scalar quantity $a_k^t a_k$ appearing in the equation in Step 2 of the algorithm is simply the number of pixels in which the kth beam passes through the center. Likewise, notice that the scalar quantity

$b_k - \mathbf{a}_k^t \mathbf{x}_{k-1}^{(p)}$ in that same equation can be interpreted as the *excess kth beam density* that results if the pixel densities are set equal to the entries of $\mathbf{x}_{k-1}^{(p)}$. This provides the following interpretation of our ART iteration scheme for the center-of-pixel method:

> *Generate the pixel densities of each iterate by distributing the excess beam density of successive beams in the scan evenly among those pixels in which the beam passes through the center. When the last beam in the scan has been reached, return to the first beam and continue.*

EXAMPLE 19.2

We shall use Algorithm 19.2 to find the unknown pixel densities of the nine pixels arranged in the 3×3 array illustrated in Figure 19.10. These nine pixels are scanned using the parallel mode with twelve beams whose measured beam densities are indicated in the figure. We choose the center-of-pixel method to set up the twelve beam equations. (In Exercises 19.7 and 19.8 the reader is asked to set up the beam equations using the center line and area methods.) As the reader can verify, the beam equations are

$$x_7 + x_8 + x_9 = 13.00$$

$$x_4 + x_5 + x_6 = 15.00$$

$$x_1 + x_2 + x_3 = 8.00$$

$$x_6 + x_8 + x_9 = 14.79$$

$$x_3 + x_5 + x_7 = 14.31$$

$$x_1 + x_2 + x_4 = 3.81$$

$$x_3 + x_6 + x_9 = 18.00$$

$$x_2 + x_5 + x_8 = 12.00$$

$$x_1 + x_4 + x_7 = 6.00$$

$$x_2 + x_3 + x_6 = 10.51$$

$$x_1 + x_5 + x_9 = 16.13$$

$$x_4 + x_7 + x_8 = 7.04$$

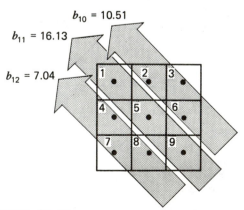

Figure 19.10

	Pixel Densities								
	x_1	x_2	x_3	x_4	x_5	x_6	x_7	x_8	x_9
$x_0^{(1)}$	.00	.00	.00	.00	.00	.00	.00	.00	.00
$x_1^{(1)}$	.00	.00	.00	.00	.00	.00	4.33	4.33	4.33
$x_2^{(1)}$	.00	.00	.00	5.00	5.00	5.00	4.33	4.33	4.33
$x_3^{(1)}$	2.67	2.67	2.67	5.00	5.00	5.00	4.33	4.33	4.33
$x_4^{(1)}$	2.67	2.67	2.67	5.00	5.00	5.37	4.33	4.71	4.71
$x_5^{(1)}$	2.67	2.67	3.44	5.00	5.77	5.37	5.10	4.71	4.71
$x_6^{(1)}$	.49	.49	3.44	2.83	5.77	5.37	5.10	4.71	4.71
$x_7^{(1)}$	.49	.49	4.93	2.83	5.77	6.87	5.10	4.71	6.20
$x_8^{(1)}$	.49	.84	4.93	2.83	6.11	6.87	5.10	5.05	6.20
$x_9^{(1)}$	-.31	.84	4.93	2.02	6.11	6.87	4.30	5.05	6.20
$x_{10}^{(1)}$	-.31	.13	4.22	2.02	6.11	6.16	4.30	5.05	6.20
$x_{11}^{(1)}$	1.06	.13	4.22	2.02	7.49	6.16	4.30	5.05	7.58
$x_{12}^{(1)}$	1.06	.13	4.22	.58	7.49	6.16	2.85	3.61	7.58
$x_{12}^{(2)}$	2.03	.69	4.42	1.34	7.49	5.39	2.65	3.04	6.61
$x_{12}^{(3)}$	1.78	.51	4.52	1.26	7.49	5.48	2.56	3.22	6.86
$x_{12}^{(4)}$	1.82	.52	4.62	1.37	7.49	5.37	2.45	3.22	6.82
$x_{12}^{(5)}$	1.79	.49	4.71	1.43	7.49	5.31	2.37	3.25	6.85
$x_{12}^{(10)}$	1.68	.44	5.03	1.70	7.49	5.03	2.04	3.29	6.96
$x_{12}^{(20)}$	1.49	.48	5.29	2.00	7.49	4.73	1.79	3.25	7.15
$x_{12}^{(30)}$	1.38	.55	5.34	2.11	7.49	4.62	1.74	3.19	7.26
$x_{12}^{(40)}$	1.33	.59	5.33	2.14	7.49	4.59	1.75	3.15	7.31
$x_{12}^{(45)}$	1.32	.60	5.32	2.15	7.49	4.59	1.76	3.14	7.32

First Cycle of Iterates

Table 19.2

Table 19.2 illustrates the results of the iteration scheme starting with an initial iterate $x_0^{(1)} = 0$. The table gives the values of each of the first cycle of iterates, $x_1^{(1)}$ through $x_{12}^{(1)}$, but thereafter only gives the iterates $x_{12}^{(p)}$ for various values of p. The iterates $x_{12}^{(p)}$ start repeating to two decimal places for $p \geq 45$, and so we take the entries of $x_{12}^{(45)}$ as approximate values of the nine pixel densities.

CLOSING REMARKS

We close this chapter by noting that the field of computed tomography is presently a very active research area. In fact, the ART scheme discussed in this chapter has been replaced in commercial systems by more sophisticated techniques that are faster and provide a more accurate view of the cross-section. However, all of the new techniques address the same basic mathematical problem: finding a good approximate solution of a large overdetermined inconsistant linear system of equations.

Readers interested in pursuing computed tomography further might find the following references useful:

[1] Richard Gordon, Gabor T. Herman, and Steven A. Johnson, "Image reconstruction from projections", *Scientific American*, October, 1975, pages 56 - 68.

[2] Gabor T. Herman, *Image Reconstruction from Projections - The Fundamentals of Computerized Tomography*, Academic Press, 1980.

[3] L.A. Shepp and J.B. Kruskal, "Computerized Tomography: The new medical x-ray technology", *American Mathematical Monthly*, Volume 85, June-July 1978, pages 420 - 439.

[4] Richard Gordon, "A Tutorial on ART", *IEEE Transactions on Nuclear Science*, Volume NS-21, June 1974, pages 78 - 93.

EXERCISES

19.1 (a) Setting $x_k^{(p)} = \left(x_{k1}^{(p)}, \ x_{k2}^{(p)} \right)$, show that the three projection equations

$$x_k^{(p)} = x_{k-1}^{(p)} + \frac{\left(b_k - \mathbf{a}_k^t x_{k-1}^{(p)} \right)}{\mathbf{a}_k^t \mathbf{a}_k} \mathbf{a}_k \ , \qquad k = 1, 2, 3$$

for the three lines in Eq. (19.5) can be written as

$k = 1:$

$$x_{11}^{(p)} = \frac{1}{2} \left[2 + x_{01}^{(p)} - x_{02}^{(p)} \right]$$

$$x_{12}^{(p)} = \frac{1}{2} \left[2 - x_{01}^{(p)} + x_{02}^{(p)} \right]$$

$k = 2:$

$$x_{21}^{(p)} = \frac{1}{5} \left[-2 + 4x_{11}^{(p)} + 2x_{12}^{(p)} \right]$$

$$x_{22}^{(p)} = \frac{1}{5} \left[4 + 2x_{11}^{(p)} + x_{12}^{(p)} \right]$$

$k = 3:$

$$x_{31}^{(p)} = \frac{1}{10} \left[9 + x_{21}^{(p)} + 3x_{22}^{(p)} \right]$$

$$x_{32}^{(p)} = \frac{1}{10} \left[-3 + 3x_{21}^{(p)} + 9x_{22}^{(p)} \right]$$

where $\left(x_{01}^{(p+1)}, \ x_{02}^{(p+1)} \right) = \left(x_{31}^{(p)}, \ x_{32}^{(p)} \right)$ for $p = 1, 2, \ \ldots \ .$

(b) Show that the three pairs of equations in part (a) can be combined to produce

$$x_{31}^{(p)} = \frac{1}{20} \left[28 + x_{31}^{(p-1)} - x_{32}^{(p-1)} \right]$$

$$x_{32}^{(p)} = \frac{1}{20} \left[24 + 3x_{31}^{(p-1)} - 3x_{32}^{(p-1)} \right]$$

$p = 1, 2, \ \ldots$

where $\left(x_{31}^{(0)}, \ x_{32}^{(0)} \right) = \left(x_{01}^{(1)}, \ x_{02}^{(1)} \right) = x_1^{(0)} .$

Remark: Using this pair of equations, one complete cycle of three orthogonal projections can be performed in a single step.

(c) Since $x_3^{(p)}$ tends to the limit point x_3^* as $p \to \infty$, the equations in part (b) become

$$x_{31}^* = \frac{1}{20}\left[28 + x_{31}^* - x_{32}^*\right]$$

$$x_{32}^* = \frac{1}{20}\left[24 + 3x_{31}^* - 3x_{32}^*\right]$$

as $p \to \infty$. Solve this linear system for $x_3^* = (x_{31}^*, x_{32}^*)$.

Remark: The simplifications of the ART formulas described in this exercise are impractical for the large linear systems that arise in realistic computed tomography problems.

19.2 Use the result of Exercise 19.1(b) to find $x_3^{(1)}$, $x_3^{(2)}$, ..., $x_3^{(6)}$ to five decimal places in Example 19.2 using the following initial points:

(a) $x_0^{(1)} = (0, 0)$ (b) $x_0^{(1)} = (1, 1)$ (c) $x_0^{(1)} = (148, -15)$.

19.3 (a) Show directly that the points of the limit cycle in Example 19.1,

$$x_1^* = (12/11, 10/11), \quad x_2^* = (46/55, 78/55), \quad x_3^* = (31/22, 27/22),$$

form a triangle whose vertices lie on the lines L_1, L_2, and L_3 and whose sides are perpendicular to these lines (Fig. 19.8(c)).

(b) Using the equations derived in Exercise 19.1(a), show that if $x_0^{(1)} = x_3^* = (31/22, 27/22)$, then

$$x_1^{(1)} = x_1^* = (12/11, 10/11)$$

$$x_2^{(1)} = x_2^* = (46/55, 78/55)$$

$$x_3^{(1)} = x_3^* = (31/22, 27/22).$$

Remark: Either part of this exercise shows that successive orthogonal projections of any point on the limit cycle will move around the limit cycle indefinitely.

19.4 The following three lines in the $x_1 x_2$-plane,

$$
\begin{aligned}
L_1: &\quad x_2 = 1 \\
L_2: &\quad x_1 - x_2 = 2 \\
L_3: &\quad x_1 - x_2 = 0
\end{aligned}
$$

do not have a common intersection. Draw an accurate sketch of the three lines and graphically perform several cycles of the orthogonal projections described in Algorithm 19.1 beginning with the initial point $x_1^{(0)} = (0, 0)$. On the basis of your sketch, determine the three points of the limit cycle.

19.5 Prove Theorem 19.1 by verifying that

(a) the point x_p as defined in the theorem lies on the line $a^t x = b$ (i.e. $a^t x_p = b$), and

(b) the vector $x_p - x^*$ is orthogonal to the line $a^t x = b$ (i.e. $x_p - x^*$ is parallel to a).

19.6 As stated in the text, the iterates $x_M^{(1)}, x_M^{(2)}, x_M^{(3)}, \ldots$ defined in Algorithm 19.2 will converge to a unique limit point x_M^* if the vectors $a_1, a_2, \ldots, a_M$ span R^N. Show that if this is the case and if the center-of-pixel method is used, then the center of each of the N pixels in the field of view is crossed by at least one of the M beams in the scan.

19.7 Construct the twelve beam equations in Example 19.2 using the center line method. Assume that the distance between the center lines of adjacent beams is equal to the width of a single pixel.

19.8 Construct the twelve beam equations in Example 19.2 using the area method. Assume that the width of each beam is equal to the width of a single pixel and that the distance between the center lines of adjacent beams is also equal to the width of a single pixel.

Linear Programming 1: A Geometric Approach

A geometric technique for maximizing or mini-
mizing a linear expression in two variables
subject to a set of linear constraints is de-
scribed.

PREREQUISITES: Linear systems
                       Linear inequalities

INTRODUCTION

The study of linear programming theory has expanded greatly since the pioneer work of George Dantzig in the late nineteen-for-ties. Today, linear programming is applied to a wide variety of problems in industry and science. In this chapter we present a geo-metric approach to the solution of simple linear programming prob-lems. In Chapters 21 and 22 we develop the algebraic theory re-quired to solve more general problems in this field.
Let us begin with some examples:

EXAMPLE 20.1 A candy manufacturer has 130 pounds of chocolate-covered cherries and 170 pounds of chocolate-covered mints in stock. He decides to sell them in the form of two different mixtures. One mixture will contain half cherries and half mints and will sell for $2.00 per pound. The other mixture will contain one-third cherries and two-thirds mints and will sell for $1.25 per pound. How many pounds of each mixture should the candy manufacturer prepare in order to maximize his sales revenue?

Let us first formulate this problem mathematically. Let the mixture of half cherries and half mints be called mix A, and let x_1 be the number of pounds of this mixture to be prepared. Let the mixture of one-third cherries and two-thirds mints be called mix B, and let x_2 be the number of pounds of this mixture to be prepared. Since mix A sells for $2.00 per pound and mix B sells for $1.25 per pound, the total sales z (in dollars) will be

$$z = 2.00x_1 + 1.25x_2.$$

Since each pound of mix A contains 1/2 pound of cherries and each pound of mix B contains 1/3 pound of cherries, the total number of pounds of cherries used in both mixtures is

$$\frac{1}{2}x_1 + \frac{1}{3}x_2.$$

Similarly, since each pound of mix A contains 1/2 pound of mints and each pound of mix B contains 2/3 pound of mints, the total number of pounds of mints used in both mixtures is

$$\frac{1}{2}x_1 + \frac{2}{3}x_2.$$

Because the manufacturer can use at most 130 pounds of cherries and 170 pounds of mints, we must have

$$\frac{1}{2}x_1 + \frac{1}{3}x_2 \leq 130$$

$$\frac{1}{2}x_1 + \frac{2}{3}x_2 \leq 170.$$

Also, since x_1 and x_2 cannot be negative numbers, we must have

$$x_1 \geq 0 \qquad \text{and} \qquad x_2 \geq 0.$$

The problem can therefore be formulated mathematically as follows:

Find values of x_1 *and* x_2 *which maximize*

$$z = 2.00x_1 + 1.25x_2$$

subject to

$$\frac{1}{2}x_1 + \frac{1}{3}x_2 \leq 130$$

$$\frac{1}{2}x_1 + \frac{2}{3}x_2 \leq 170$$

$$x_1 \geq 0$$

$$x_2 \geq 0.$$

In the next section we shall show how to solve this type of mathematical problem geometrically.

EXAMPLE 20.2

A woman has up to $10,000 to invest. Her broker suggests investing in two bonds, A and B. Bond A is a rather risky bond with an annual yield of 10%, and bond B is a rather safe bond with an annual yield of 7%. After some consideration, she decides to invest at most $6,000 in bond A, at least $2,000 in bond B, and to invest at least as much in bond A as in bond B. How should she invest her $10,000 in order to maximize her annual yield?

In order to formulate this problem mathematically, let x_1 be the number of dollars to be invested in bond A and let x_2 be the number of dollars to be invested in bond B. Since each dollar invested in bond A earns $.10 per year and each dollar invested in bond B earns $.07 per year, the total dollar amount z earned each year by both bonds is

$$z = .10x_1 + .07x_2.$$

The constraints imposed can be formulated mathematically as follows:

Invest no more than $10,000:	$x_1 + x_2 \leq 10,000$
Invest at most $6,000 in bond A:	$x_1 \leq 6,000$
Invest at least $2,000 in bond B:	$x_2 \geq 2,000$
Invest at least as much in bond A as in bond B:	$x_1 \geq x_2.$

We also have the implicit assumption that x_1 and x_2 are nonnegative:

$$x_1 \geq 0 \quad \text{and} \quad x_2 \geq 0.$$

Thus, the complete mathematical formulation of the problem is as follows:

Find values of x_1 and x_2 which maximize

$$z = .10x_1 + .07x_2$$

subject to

$$x_1 + x_2 \leq 10,000$$

$$x_1 \leq 6,000$$

$$x_2 \geq 2,000$$

$$x_1 - x_2 \geq 0$$

$$x_1 \geq 0$$

$$x_2 \geq 0.$$

EXAMPLE 20.3

A student desires to design a breakfast of corn flakes and milk which is as economical as possible. On the basis of what he eats during his other meals, he decides that his breakfast should supply him with at least nine grams of protein, at least one-third the recommended daily allowance (RDA) of vitamin D, and at least one-fourth the RDA of calcium. He finds the following nutrition information on the milk and corn flakes containers:

	Milk ½ cup	Corn Flakes 1 ounce
Cost	7.5¢	5.0¢
Protein	4 grams	2 grams
Vitamin D	1/8 of RDA	1/10 of RDA
Calcium	1/6 of RDA	none

In order not to have his mixture too soggy or too dry, the student decides to limit himself to mixtures which contain one to three ounces of corn flakes per cup of milk, inclusive. What quantities of milk and corn flakes should he use to minimize the cost of his breakfast?

For the mathematical formulation of this problem, let x_1 be the quantity of milk used measured in ½-cup units and let x_2 be the quantity of corn flakes used measured in 1-ounce units. Then if z is the cost of the breakfast in cents, we may write the following:

Cost of breakfast: $\qquad z = 7.5x_1 + 5.0x_2$

At least nine grams of protein: $\qquad 4x_1 + 2x_2 \geq 9$

At least 1/3 of RDA of vitamin D: $\qquad \frac{1}{8}x_1 + \frac{1}{10}x_2 \geq \frac{1}{3}$

At least 1/4 of RDA of calcium: $\frac{1}{6} x_1 \geq \frac{1}{4}$

At least one ounce of corn flakes
per cup (two ½-cups) of milk: $x_2/x_1 \geq \frac{1}{2}$ (or $x_1 - 2x_2 \leq 0$)

At most three ounces of corn
flakes per cup (two ½-cups) of milk: $x_2/x_1 \leq \frac{3}{2}$ (or $3x_1 - 2x_2 \geq 0$)

As before, we also have the implicit assumption that $x_1 \geq 0$ and
$x_2 \geq 0$. Thus the complete mathematical formulation of the problem
is as follows:

Find values of x_1 and x_2 which minimize

$$z = 7.5x_1 + 5.0x_2$$

subject to

$$4x_1 + 2x_2 \geq 9$$

$$\frac{1}{8} x_1 + \frac{1}{10} x_2 \geq \frac{1}{3}$$

$$\frac{1}{6} x_1 \geq \frac{1}{4}$$

$$x_1 - 2x_2 \leq 0$$

$$3x_1 - 2x_2 \geq 0$$

$$x_1 \geq 0$$

$$x_2 \geq 0.$$

GEOMETRIC SOLUTION OF LINEAR PROGRAMMING PROBLEMS

Each of the three examples in the introduction is a special
case of the following problem:

PROBLEM 20.1 *Find values of* x_1 *and* x_2 *which either maximize or minimize*

$$z = c_1 x_2 + c_2 x_2 \qquad (20.1)$$

subject to

$$a_{11}x_1 + a_{12}x_2 \ (\leq)(\geq)(=) \ b_1$$
$$a_{21}x_1 + a_{22}x_2 \ (\leq)(\geq)(=) \ b_2$$
$$\vdots \qquad\qquad (20.2)$$
$$a_{m1}x_1 + a_{m2}x_2 \ (\leq)(\geq)(=) \ b_m$$

and

$$x_1 \geq 0, \qquad x_2 \geq 0. \qquad (20.3)$$

In each of the m equations of (20.2), any one of the symbols $\leq$, $\geq$, $=$ may be used.

Problem 17.1 is the *general linear programming problem* in two variables. The linear function z in (20.1) is called the *objective function*. Equations (20.2) and (20.3) are called the *constraints*; in particular, Eqs. (20.3) are called the *nonnegativity constraints* on the variables x_1 and x_2.

We shall now show how to solve a linear programming problem in two variables graphically. A pair of values (x_1, x_2) which satisfy all of the constraints is called a *feasible solution*. The set of all feasible solutions determines a subset of the $x_1 x_2$-plane called the *feasible region*. Our desire is to find a feasible solution which maximizes the objective function. Such a solution is called an *optimal solution*.

To examine the feasible region of a linear programming problem, let us note that each constraint of the form

$$a_{i1}x_1 + a_{i2}x_2 = b_i$$

defines a line in the $x_1 x_2$-plane, while each constraint of the form

$$a_{i1}x_1 + a_{i2}x_2 \leq b_i$$

or

$$a_{i1}x_1 + a_{i2}x_2 \geq b_i$$

defines a half-plane which includes its boundary line

$$a_{i1}x_1 + a_{i2}x_2 = b_i.$$

Thus, the feasible region is always an intersection of finitely many lines and half-planes. For example, the four constraints

$$\frac{1}{2}x_1 + \frac{1}{3}x_2 \le 130$$

$$\frac{1}{2}x_1 + \frac{2}{3}x_2 \le 170$$

$$x_1 \ge 0$$

$$x_2 \ge 0$$

of Example 20.1 define the half-planes illustrated in Fig. 20.1(a), (b), (c), and (d). The feasible region of this problem is thus the intersection of these four half-planes, which is illustrated in Fig. 20.1(e).

It can be shown that the feasible region of a linear programming problem has a boundary consisting of a finite number of straight-line segments. If the feasible region can be enclosed in a sufficiently large circle, it is called *bounded* (Fig. 20.1(e)); otherwise it is called *unbounded* (Fig. 20.5). If the feasible region is *empty* (contains no points), then the constraints are inconsistent and the linear programming problem has no solution (Fig. 20.6).

Those boundary points of a feasible region which are intersections of two of the straight-line boundary segments are called *extreme points*. (They are also called *corner points* or *vertex points*.) For example, from Fig. 20.1(e), the feasible region of Example 20.1 has four extreme points:

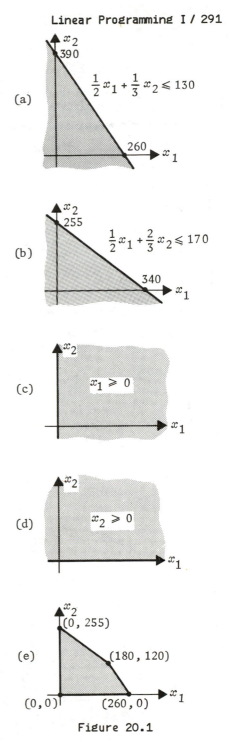

Figure 20.1

$$(0, 0), \quad (0, 255), \quad (180, 120), \quad (260, 0). \tag{20.4}$$

The importance of the extreme points of a feasible region is shown by the following theorem:

THEOREM 20.1 *If the feasible region of a linear programming problem is nonempty and bounded, then the objective function attains both a maximum and minimum value and these occur at extreme points of the feasible region. If the feasible region is unbounded, then the objective function may or may not attain a maximum or minimum value; however, if it attains a maximum or minimum value, it does so at an extreme point.*

Figure 20.2 suggests the idea behind the proof of this theorem. Since the objective function

$$z = c_1 x_1 + c_2 x_2$$

of a linear programming problem is a linear function of x_1 and x_2, its level curves (the curves along which z has constant values) are straight lines. As we move in a direction perpendicular to these level curves, the objective function either increases or decreases monotonically. Within a bounded feasible region, the maximum and minimum values of z must therefore occur at extreme points, as Figure 20.2 indicates.

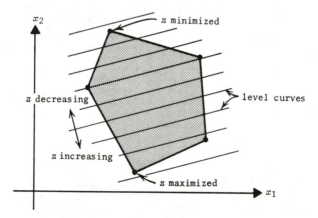

Figure 20.2

In the next few examples, we use Theorem 20.1 to solve several linear programming problems and illustrate the variations in the nature of the solutions which may occur.

EXAMPLE 20.1 (REVISITED) From Figure 20.1(e), we see that the feasible region of Example 20.1 is bounded. Consequently, from Theorem 20.1 the objective function

$$z = 2.00x_1 + 1.25x_2$$

attains both its minimum and maximum values at extreme points. The four extreme points and the corresponding values of z are given in the following table

Extreme point (x_1, x_2)	Value of $z = 2.00x_1 + 1.25x_2$
(0, 0)	0
(0, 255)	318.75
(180, 120)	510.00
(260, 0)	520.00

We see that the largest value of z is 520.00, and the corresponding optimal solution is (260, 0). Thus, the candy manufacturer attains maximum sales of $520 when he produces 260 pounds of mixture A and none of mixture B.

EXAMPLE 20.4 Find values of x_1 and x_2 which maximize

$$z = x_1 + 3x_2$$

subject to

$$2x_1 + 3x_2 \leq 24$$
$$x_1 - x_2 \leq 7$$
$$x_2 \leq 6$$
$$x_1 \geq 0$$
$$x_2 \geq 0.$$

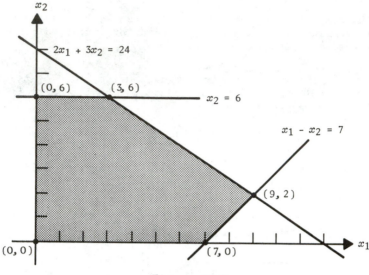

Figure 20.3

SOLUTION In Fig. 20.3 we have drawn the feasible region of this problem. Since it is bounded, the maximum value of z is attained at one of the five extreme points. The values of the objective function at the five extreme points are given in the following table:

Extreme point (x_1, x_2)	Value of $z = x_1 + 3x_2$
(0, 6)	18
(3, 6)	21
(9, 2)	15
(7, 0)	7
(0, 0)	0

From this table, the maximum value of z is 21, which is attained at $x_1 = 3$ and $x_2 = 6$.

EXAMPLE 20.5 Find values of x_1 and x_2 which maximize

$$z = 4x_1 + 6x_2$$

subject to

$$2x_1 + 3x_2 \leq 24$$

$$x_1 - x_2 \leq 7$$

$$x_2 \leq 6$$

$$x_1 \geq 0$$

$$x_2 \geq 0.$$

SOLUTION The constraints in this problem are identical to the constraints in Example 20.4, and so the feasible region of this problem is also given by Fig. 20.3. The values of the objective function at the extreme points are as follows:

Extreme point (x_1, x_2)	Value of $z = 4x_1 + 6x_2$
(0, 6)	36
(3, 6)	48
(9, 2)	48
(7, 0)	28
(0, 0)	0

We see that the objective function attains a maximum value of 48 at two adjacent extreme points, (3, 6) and (9, 2). This shows that an optimal solution to a linear programming problem need not be unique. As we ask the reader to show in Exercise 20.9, if the objective function has the same value at two adjacent extreme points, it has the same value at all points on the straight-line boundary segment connecting the two extreme points. Thus, in this example the maximum value of z is attained at all points on the straight-line segment connecting the extreme points (3, 6) and (9, 2).

EXAMPLE 20.6 Find values of x_1 and x_2 which minimize

$$z = 2x_1 - x_2$$

subject to

$$2x_1 + 3x_2 = 12$$

$$2x_1 - 3x_2 \geq 0$$

$$x_1 \geq 0$$

$$x_2 \geq 0.$$

SOLUTION In Fig. 20.4 we have drawn the feasible region of this problem. Because one of the constraints is an equality constraint, the feasible region is a straight line segment with two extreme points. The values of z at the two extreme points are as follows:

Extreme point (x_1, x_2)	Value of $z = 2x_1 - x_2$
(3, 2)	4
(6, 0)	12

The minimum value of z is thus 4 and is attained at $x_1 = 3$ and $x_2 = 2$.

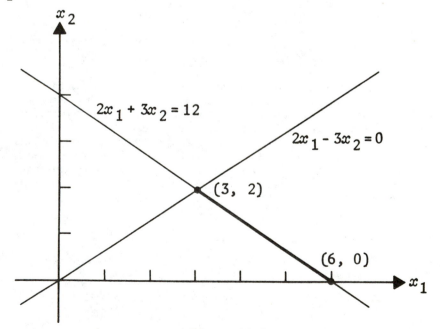

Figure 20.4

EXAMPLE 20.7 Find values of x_1 and x_2 which maximize

$$z = 2x_1 + 5x_2$$

subject to

$$2x_1 + x_2 \geq 8$$

$$-4x_1 + x_2 \leq 2$$

$$2x_1 - 3x_2 \leq 0$$

$$x_1 \geq 0$$

$$x_2 \geq 0.$$

SOLUTION The feasible region of this linear programming problem is illustrated in Fig. 20.5. Since it is unbounded, we are not assured by Theorem 20.1 that the objective function attains a maximum value. In fact, it is easily seen that since the feasible region contains points for which both x_1 and x_2 are arbitrarily large and positive, then the objective function

$$z = 2x_1 + 5x_2$$

can be made arbitrarily large and positive. This problem has no optimal solution. Instead, we say the problem has an *unbounded solution*.

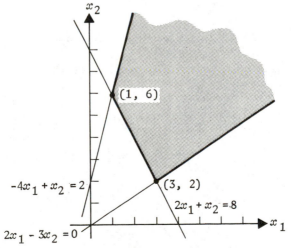

Figure 20.5

EXAMPLE 20.8 Find values of x_1 and x_2 which maximize

$$z = -5x_1 + x_2$$

subject to

$$2x_1 + x_2 \geq 8$$

$$-4x_1 + x_2 \leq 2$$

$$2x_1 - 3x_2 \leq 0$$

$$x_1 \geq 0$$

$$x_2 \geq 0.$$

SOLUTION The above constraints are the same as those in Example 20.7, so that the feasible region of this problem is also given by Fig. 20.5. In Exercise 20.10, we ask the reader to show that the objective function of this problem attains a maximum within the feasible region. By Theorem 20.1, this maximum must be attained at an extreme point. The values of z at the two extreme points of the feasible region are given by

Extreme point (x_1, x_2)	Value of $z = -5x_1 + x_2$
(1, 6)	1
(3, 2)	-13

The maximum value of z is thus 1 and is attained at the extreme point $x_1 = 1$, $x_2 = 6$.

EXAMPLE 20.9 Find the values x_1 and x_2 which minimize

$$z = 3x_1 - 8x_2$$

subject to

$$2x_1 - x_2 \leq 4$$

$$3x_1 + 11x_2 \leq 33$$

$$3x_1 + 4x_2 \geq 24$$

$$x_1 \geq 0$$

$$x_2 \geq 0.$$

SOLUTION As can be seen from Fig. 20.6, the intersection of the five half-planes defined by the five constraints is empty. This linear programming problem has no feasible solutions since the constraints are inconsistant.

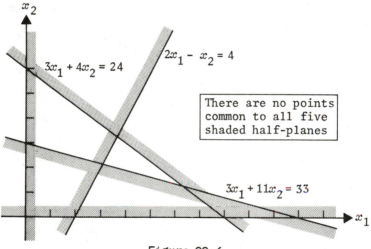

Figure 20.6

EXERCISES

20.1 Find values of x_1 and x_2 which maximize

$$z = 3x_1 + 2x_2$$

subject to

$$2x_1 + 3x_2 \leq 6$$

$$2x_1 - x_2 \geq 0$$

$$x_1 \leq 2$$

$$x_2 \leq 1$$

$$x_1 \geq 0$$

$$x_2 \geq 0.$$

20.2 Find values of x_1 and x_2 which minimize

$$z = 3x_1 - 5x_2$$

subject to

$$2x_1 - x_2 \leq -2$$

$$4x_1 - x_2 \geq 0$$

$$x_2 \leq 3$$

$$x_1 \geq 0$$

$$x_2 \geq 0.$$

20.3 Find values of x_1 and x_2 which minimize

$$z = -3x_1 + 2x_2$$

subject to

$$3x_1 - x_2 \geq -5$$

$$-x_1 + x_2 \geq 1$$

$$2x_1 + 4x_2 \geq 12$$

$$x_1 \geq 0$$

$$x_2 \geq 0.$$

20.4 Solve the linear programming problem posed in Example 20.2.

20.5 Solve the linear programming problem posed in Example 20.3.

20.6 A trucking firm ships the containers of two companies, A and B. Each container of Company A weighs 40 pounds and is 2 cubic feet in volume. Each container of Company B weighs 50 pounds and is 3 cubic feet in volume. The trucking firm charges Company A \$2.20 for each container shipped and charges Company B \$3.00 for each container shipped. If one of the firm's trucks cannot carry more than 37,000 pounds and cannot hold more than 2000 cubic feet, how many containers from companies A and B should a truck carry to maximize the shipping charges?

20.7 Repeat Exercise 20.6 if the trucking firm raises its price for shipping a container of Company A to \$2.50.

20.8 A manufacturer produces sacks of chicken feed from two ingredients, A and B. Each sack is to contain at least 10 ounces of nutrient N_1, at least 8 ounces of nutrient N_2, and at least 12 ounces of nutrient N_3. Each pound of ingredient A contains 2 ounces of nutrient N_1, 2 ounces of nutrient N_2, and 6 ounces of nutrient N_3. Each pound of ingredient B contains 5 ounces of nutrient N_1. 3 ounces of nutrient N_2, and 4 ounces of nutrient N_3. If ingredient A costs 8¢ per pound and ingredient B costs 9¢ per pound, how much of each ingredient should the manufacturer use in each sack of feed to minimize his costs?

20.9 If the objective function of a linear programming problem has the same value at two adjacent extreme points, show that it has the same value at all points on the straight-line segment connecting the two extreme points. Hint: if (x_1', x_2') and (x_1'', x_2'') are any two points in the plane, a point (x_1, x_2) lies on the straight-line segment connecting them if

$$x_1 = tx_1' + (1 - t)x_1''$$

and

$$x_2 = tx_2' + (1 - t)x_2''$$

where t is a number in the interval $[0, 1]$.

20.10 Show that the objective function in Example 20.8 attains a maximum value in the feasible set. Hint: Examine the level curves of the objective function.

Linear Programming 2: Basic Concepts

The basic concepts needed to develop the simplex method for solving linear programming problems are presented.

PREREQUISITES: Linear Systems
Matrices
Linear independence
Euclidean space R^n
Chapter 20: Linear Programming 1

INTRODUCTION

In the last chapter we presented a geometric technique for solving linear programming problems in two variables. However, this technique is not practical for the solution of linear programming problems in three or more variables. In this chapter we develop the fundamental ideas behind an algebraic technique, called the *simplex method*, for solving linear programming problems in any number of variables. The simplex method itself is presented in Chapter 22.

The *general linear programming problem in n variables*, described below, is analogous to Problem 20.1 in the last chapter.

PROBLEM 21.1 *Find values of* $x_1, x_2, \ldots, x_n$ *which either maximize or minimize*

$$z = c_1 x_1 + c_2 x_2 + \cdots + c_n x_n \qquad (21.1)$$

subject to

$$a_{11} x_1 + a_{12} x_2 + \cdots + a_{1n} x_n \ (\leq)(\geq)(=) \ b_1$$
$$a_{21} x_1 + a_{22} x_2 + \cdots + a_{2n} x_n \ (\leq)(\geq)(=) \ b_2$$
$$\vdots \qquad\qquad (21.2)$$
$$a_{m1} x_1 + a_{m2} x_2 + \cdots + a_{mn} x_n \ (\leq)(\geq)(=) \ b_m$$

and

$$x_i \geq 0, \qquad \text{for } i = 1, 2, \ldots, n. \qquad (21.3)$$

As in Chapter 20, the linear function z in (21.1) is called the *objective function* and conditions (21.2) and (21.3) are called the *constraints* of the problem.

For our purposes, it is necessary that we first convert our general linear programming problem to the following *standard form*:

PROBLEM 21.2 *Find values of* $x_1, x_2, \ldots, x_n$ *which maximize*

$$z = c_1 x_1 + c_2 x_2 + \cdots + c_n x_n \qquad (21.4)$$

subject to

$$a_{11} x_1 + a_{12} x_2 + \cdots + a_{1n} x_n = b_1$$
$$a_{21} x_1 + a_{22} x_2 + \cdots + a_{2n} x_n = b_2$$
$$\vdots \qquad\qquad (21.5)$$
$$a_{m1} x_1 + a_{m2} x_2 + \cdots + a_{mn} x_n = b_m$$

and

$$x_i \geq 0, \qquad \text{for } i = 1, 2, \ldots, n. \qquad (21.6)$$

Any linear programming problem can always be put in this standard form using the following three steps:

Step 1. Convert a minimization problem to a maximization problem by defining a new objective function

$$z' = -z.$$

For example, the problem of minimizing the objective function

$$z = 2x_1 + 3x_2 - 5x_3$$

is equivalent to the problem of maximizing the objective function

$$z' = -2x_1 - 3x_2 + 5x_3 .$$

Step 2. Convert a $\geq$ constraint to a $\leq$ constraint by multiplying the inequality by -1. Thus, the constraint

$$-x_1 + 2x_2 - 4x_3 \geq 6$$

is the same as the constraint

$$x_1 - 2x_2 + 4x_3 \leq -6.$$

Step 3. Convert a $\leq$ constraint to an equality constraint by adding a nonnegative *slack variable* to the lefthand side of the inequality. For example, if the original problem contains three variables, and one of the constraints is

$$x_1 - 2x_2 + 4x_3 \leq -6,$$

we add a new variable $x_4 \geq 0$ to the lefthand side to obtain

$$x_1 - 2x_2 + 4x_3 + x_4 = -6.$$

The variable x_4 takes up the slack between the two sides of the inequality. In this way, a new variable is added to the problem for each $\leq$ constraint. We assign each slack variable introduced a coefficient $c_i = 0$ in the objective function, so that the objective function is not affected by the values of the slack variables.

EXAMPLE 21.1
Convert the following linear programming problem to one in standard form:

$$\text{Minimize} \quad z = 3x_1 - 2x_2 + x_3 - x_4$$

Subject to

$$2x_1 + 5x_2 - 6x_3 - x_4 \leq 2$$

$$x_1 - 7x_2 - 5x_3 + 2x_4 \geq 6$$

$$2x_1 - 8x_2 - 8x_3 + 6x_4 = 5$$

and

$$x_1, x_2, x_3, x_4 \geq 0.$$

SOLUTION The first step is to multiply the objective function by -1 to convert the problem to a maximization problem. The second step is to multiply the second constraint by -1 to convert it to a $\leq$ inequality. The third step is to add slack variables x_5 and x_6 to the first and second constraints to convert them to equalities. The final problem in standard form is

$$\text{Maximize} \quad z' = -3x_1 + 2x_2 - x_3 + x_4 + 0x_5 + 0x_6$$

subject to

$$2x_1 + 5x_2 - 6x_3 - x_4 + x_5 \qquad\quad = 2$$

$$-x_1 + 7x_2 + 5x_3 - 2x_4 \qquad + x_6 = -6$$

$$2x_1 - 8x_2 - 8x_3 + 6x_4 \qquad\qquad = 5$$

and

$$x_1, x_2, x_3, x_4, x_5, x_6 \geq 0.$$

◆—◆—◆

It will be convenient to use matrix notation to express Problem 21.2 in a more compact form as follows (the expression $x \geq 0$ below denotes that each entry of the vector x is nonnegative):

PROBLEM 21.2 (in matrix notation)

$$\textit{Maximize} \quad z = c^t x$$

subject to

$$Ax = b$$

and

$$x \geq 0,$$

where

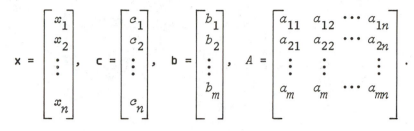

$$\mathbf{x} = \begin{bmatrix} x_1 \\ x_2 \\ \vdots \\ x_n \end{bmatrix}, \quad \mathbf{c} = \begin{bmatrix} c_1 \\ c_2 \\ \vdots \\ c_n \end{bmatrix}, \quad \mathbf{b} = \begin{bmatrix} b_1 \\ b_2 \\ \vdots \\ b_m \end{bmatrix}, \quad A = \begin{bmatrix} a_{11} & a_{12} & \cdots & a_{1n} \\ a_{21} & a_{22} & \cdots & a_{2n} \\ \vdots & \vdots & & \vdots \\ a_{m} & a_{m} & \cdots & a_{mn} \end{bmatrix}.$$

In this formulation, the problem is to find a nonnegative vector $\mathbf{x}$ in the vector space R^n which satisfies the constraint condition $A\mathbf{x} = \mathbf{b}$ and which makes the objective function $z = \mathbf{c}^t\mathbf{x}$ as large as possible. Analogous to the terminology in Chapter 20, any nonnegative vector $\mathbf{x}$ which satisfies the constraint $A\mathbf{x} = \mathbf{b}$ is called a *feasible solution* to the linear programming problem. The set of all feasible solutions in R^n is called the *feasible set* or the *feasible region* of the problem. A feasible solution which maximizes the objective function is called an *optimal solution*. As in Chapter 20, there are three possible outcomes to a linear programming problem:

(i)   The constraints are inconsistent so that there are no feasible solutions.

(ii) The feasible set is unbounded and the objective function can be made arbitrarily large.

(iii) There is at least one optimal solution.

In most realistic applications only case (iii) arises.

We now examine the nature of the feasible set of a linear programming problem. To begin, let us introduce the following definition:

DEFINITION 21.1 *A set of vectors in R^n is called* **convex** *if whenever* $\mathbf{x}_1$ *and* $\mathbf{x}_2$ *belong to the set, so does the vector*

$$\mathbf{x} = t\mathbf{x}_1 + (1 - t)\mathbf{x}_2$$

for any number t in the interval $[0, 1]$.

Geometrically, the vector $x = tx_1 +$ (1 - t)x_2 lies on the line segment connecting the tips of the vectors x_1 and x_2 (Fig. 21.1). Thus, a convex set can be viewed as one in which the line segment connecting any two points in the set also belongs to the set. Figures 21.2(a), (b), and (c) illustrate three convex sets in R^2. Figure 21.1(d) is an example of a set in R^2 which is not convex. In Exercise 21.7, we ask the reader to prove the following theorem:

(a)   Convex

(b)   Convex

```
┌─────────────────────────────────────┐
│  THEOREM 21.1  The feasible          │
│  set of a standard linear            │
│  programming problem is con-         │
│  vex.                                │
└─────────────────────────────────────┘
```

(c)   Convex

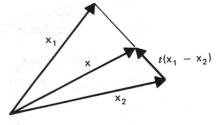

$$x = x_2 + t(x_1 - x_2)$$

$$= tx_1 + (1 - t)x_2$$

(d)   Not Convex

Figure 21.1

Figure 21.2

EXAMPLE 21.2 For the following linear programming problem:

$$\text{Maximize} \quad z = 3x_1 + x_2 - 2x_3$$

subject to

$$3x_1 + x_2 + x_3 = 10$$
$$2x_1 - x_2 + 2x_3 = 10$$

and

$$x_1, x_2, x_3 \geq 0,$$

the feasible set consists of the portion of the intersection of the two planes

$$3x_1 + x_2 + x_3 = 10$$
$$2x_1 - x_2 + 2x_3 = 10$$

which lies in the first octant of $x_1x_2x_3$-space. As we ask the reader to show in Exercise 21.8, this intersection consists of the line segment connecting the points

$(5/2, 0, 5/2)$ and $(0, 10/3, 20/3)$

(Fig. 21.3), which is clearly convex.

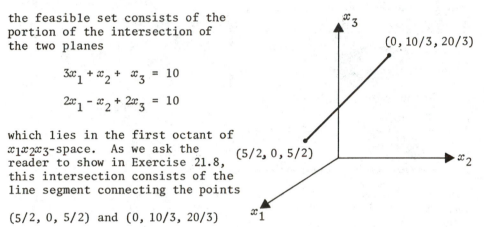

Figure 21.3

EXAMPLE 21.3 For the following linear programming problem:

$$\text{Maximize} \quad z = x_1 - 3x_2 + 2x_3$$

subject to

$$2x_1 + 4x_2 + 3x_3 = 12$$

and

$$x_1, x_2, x_3 \geq 0,$$

the feasible set consists of the portion of the plane

$$2x_1 + 4x_2 + 3x_3 = 12$$

which lies in the first octant of $x_1 x_2 x_3$-space. The reader can easily verify that Fig. 21.4 is a diagram of this feasible set and that it is convex.

●—●—●

In Chapter 20, we introduced the concept of an extreme point of a feasible set in $x_1 x_2$-space. For an arbitrary convex set in R^n, the corresponding definition is the following:

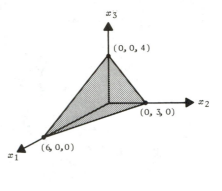

Figure 21.4

DEFINITION 21.2 *A vector* x *in a convex set is an* **extreme point** *of the convex set if*

$$x \neq \tfrac{1}{2}(x_1 + x_2)$$

for any two vectors x_1 *and* x_2 *in the set.*

In other words, the extreme points of a convex set are those points which do not lie midway between any two points in the set. For example, the extreme points of the convex set illustrated in Fig. 21.2 are the two endpoints

$$(5/2,\ 0,\ 5/2) \quad \text{and} \quad (0,\ 10/3,\ 20/3)$$

of the straight-line segment. For the convex set illustrated in Fig. 21.4, the extreme points are the three corner points

$$(6,\ 0,\ 0), \quad (0,\ 3,\ 0), \quad \text{and} \quad (0,\ 0,\ 4)$$

of the triangular-shaped region.

Recall that in the last chapter we called a set in R^2 bounded if it can be enclosed in a sufficiently large circle; that is, if there is some positive radius r such that each point $x = (x_1, x_2)$ in the set satisfies

$$\|x\| = \sqrt{x_1^2 + x_2^2} \leq r.$$

Similarly, we can define a bounded set in R^n as follows:

DEFINITION 21.3 *A set in R^n is said to be* **bounded** *if there is some positive number r such that each point* $x = (x_1, x_2, \ldots, x_n)$ *in the set satisfies*

$$\| x \| = \sqrt{x_1^2 + x_2^2 + \cdots + x_n^2} \le r .$$

The following theorem, which is just Theorem 20.1 of Chapter 20 extended to the space R^n, shows that if a linear programming problem has an optimal solution, it can be found among the extreme points of the feasible set.

THEOREM 21.2 *If the feasible set of a linear programming problem is nonempty and bounded, then the objective function attains its maximum at an extreme point of the set. If the feasible set is unbounded, then the objective function may or may not attain a maximum value; however, if it attains a maximum value, it does so at an extreme point.*

The proof of this theorem follows the same lines as that outlined in Chapter 20 for two variables. Let us apply this theorem to the two problems posed in Examples 21.2 and 21.3.

EXAMPLE 21.2 (REVISITED)

As illustrated in Fig. 21.3, the feasible set for this problem is nonempty and bounded. Consequently, by Theorem 21.2 the objective function

$$z = 3x_1 + x_2 - 2x_3$$

attains its maximum at one of the two extreme points of the set. At the extreme point

$$(5/2, 0, 5/2)$$

we have

$$z = 5/2,$$

and at the extreme point

$$(0, 10/3, 20/3)$$

we have

$$z = -10.$$

Thus, the maximum value of the objective function is $z = 5/2$, and an optimal solution to the problem is $x_1 = 5/2$, $x_2 = 0$, and $x_3 = 5/2$.

EXAMPLE 21.3 (REVISITED)
The feasible set of Example 21.3 is illustrated in Fig. 21.4. Since it is nonempty and bounded, Theorem 21.2 guarantees that the objective function

$$z = x_1 - 3x_2 + 2x_3$$

attains its maximum value at one of the three extreme points. The following table gives the values of the objective function at these extreme points:

Extreme point (x_1, x_2, x_3)	Value of $z = x_1 - 3x_2 + 2x_3$
(6, 0, 0)	6
(0, 3, 0)	-9
(0, 0, 4)	8

Thus the maximum value of z is 8 and an optimal solution is $x_1 = 0$, $x_2 = 0$, $x_3 = 4$.

In the above two examples (and in all of the examples in Chapter 20) the extreme points of the feasible sets were found geometrically. But geometric techniques are not possible if the problem has more than three variables. In such cases, we need an algebraic technique for generating the extreme points of the feasible set. In the next section we describe such an algebraic technique.

BASIC FEASIBLE SOLUTIONS

Let us reexamine the linear system

$$Ax = b \tag{21.7}$$

of Problem 21.2, where A is an $m \times n$ matrix. Although it is not essential, we will assume for simplicity that $m \leq n$; i.e., that there are no more constraints than variables. We shall also assume that the m rows of A are linearly independent. This implies that A contains m linearly independent columns since the row space and column space of any matrix have the same dimension. In Exercise 21.10, we ask the reader to show that the linear system (21.7) can also be written as

$$x_1 a_1 + x_2 a_2 + \cdots + x_n a_n = b \qquad (21.8)$$

where a_i $(i = 1, 2, \ldots, n)$ is the i-th column vector of the matrix A. Thus, solving $Ax = b$ for x is equivalent to solving the vector equation (21.8) for $x_1, x_2, \ldots, x_n$. For convenience, we say that the variable x_i "corresponds" to the column vector a_i. As noted above, there must exist at least one set of m linearly independent vectors among the column vectors $a_1, a_2, \ldots, a_n$. For example, suppose the first m of these vectors, $a_1, a_2, \ldots, a_m$, form such a set. Since these m vectors lie in R^m, and since R^m is m-dimensional, $a_1, a_2, \ldots, a_m$ constitute a basis for R^m. Thus the vector b in (21.8) is uniquely expressible as a linear combination of $a_1, a_2, \ldots, a_m$. That is, there is a unique solution of (21.8) for which

$$x_{m+1} = x_{m+2} = \cdots = x_n = 0. \qquad (21.9)$$

Similarly, any set of m linearly independent column vectors of A would lead to a solution of (21.8) in which $n - m$ of the variables are zero and the remaining m are uniquely determined. This suggests the following definition:

DEFINITION 21.4 *A vector*

$$x = \begin{bmatrix} x_1 \\ x_2 \\ \vdots \\ x_n \end{bmatrix}$$

is called a **basic solution** *of the linear system* $Ax = b$ *if* $n - m$ *of the variables* $x_1, x_2, \ldots, x_n$ *are zero and the remaining* m *variables correspond to linearly independent column vectors of* A. *The* $n - m$ *zero variables are called the* **nonbasic variables,** *and the* m *variables corresponding to the linearly independent column vectors are called the* **basic variables** *of* x.

A linear system of m equations in n variables has as many basic solutions as there are sets of m linearly independent columns among the n columns of the coefficient matrix A.

EXAMPLE 21.4 Find all basic solutions of the linear system

$$2x_1 \quad + x_3 + 4x_4 + 2x_5 = 20$$

$$x_1 + x_2 - x_3 + x_4 + 3x_5 = 10.$$

SOLUTION The coefficient matrix of the linear system is

$$A = \begin{bmatrix} 2 & 0 & 1 & 4 & 2 \\ 1 & 1 & -1 & 1 & 3 \end{bmatrix}.$$

As we can see, any two columns of this matrix are linearly independent. Thus to find a basic solution we choose any two of these columns, set the three appropriate nonbasic variables equal to zero and solve the resulting 2×2 linear system for the two basic variables. For example, if we choose the two columns

$$\mathbf{a}_1 = \begin{bmatrix} 2 \\ 1 \end{bmatrix} \quad \text{and} \quad \mathbf{a}_3 = \begin{bmatrix} 1 \\ -1 \end{bmatrix},$$

we set the nonbasic variables

$$x_2, \quad x_4, \quad \text{and} \quad x_5$$

equal to zero and obtain the system

$$2x_1 + x_3 = 20$$

$$x_1 - x_3 = 10$$

for the basic variables x_1 and x_3. This system is easily found to have the solution $x_1 = 10$ and $x_3 = 0$. The resulting basic solution to the original problem is

$$\mathbf{x} = \begin{bmatrix} 10 \\ 0 \\ 0 \\ 0 \\ 0 \end{bmatrix},$$

where we have underlined the basic variables. As the reader can verify, the ten possible pairs of columns of A lead to the following ten basic solutions:

$$\begin{bmatrix} \underline{10} \\ 0 \\ 0 \\ 0 \\ 0 \end{bmatrix} \begin{bmatrix} \underline{10} \\ 0 \\ \underline{0} \\ 0 \\ 0 \end{bmatrix} \begin{bmatrix} \underline{10} \\ 0 \\ 0 \\ \underline{0} \\ 0 \end{bmatrix} \begin{bmatrix} \underline{10} \\ 0 \\ 0 \\ 0 \\ \underline{0} \end{bmatrix} \begin{bmatrix} 0 \\ \underline{30} \\ \underline{20} \\ 0 \\ 0 \end{bmatrix} \begin{bmatrix} 0 \\ \underline{5} \\ 0 \\ \underline{5} \\ 0 \end{bmatrix} \begin{bmatrix} 0 \\ -20 \\ 0 \\ 0 \\ \underline{10} \end{bmatrix} \begin{bmatrix} 0 \\ 0 \\ -4 \\ 6 \\ 0 \end{bmatrix} \begin{bmatrix} 0 \\ 0 \\ 8 \\ 0 \\ 6 \end{bmatrix} \begin{bmatrix} 0 \\ 0 \\ 0 \\ 4 \\ 2 \end{bmatrix} \quad (21.10)$$

where we have underlined the two basic variables in each case.

◆◆◆

From (21.10), we see that the first four basic solutions are equal as vectors in the vector space R^5. Nevertheless, we shall consider them to be distinct basic solutions because they result from different pairs of linearly independent columns of A. This particular circumstance arises because one of the basic variables in each of these four basic solutions is equal to zero. In general, a basic solution is said to be *degenerate* if any of its basic variables is equal to zero. Otherwise, it is said to be *nondegenerate*.

Recall that the feasible set of a standard linear programming problem consists of those vectors which satisfy a linear equation

$$A\mathbf{x} = \mathbf{b}$$

and which satisfy the nonnegativity condition

$$\mathbf{x} \geq 0.$$

Adjoining the nonnegativity condition to the concept of a basic solution we are led to the following definition:

DEFINITION 21.5 *In a standard linear programming problem a feasible solution which is also a basic solution of the system $A\mathbf{x} = \mathbf{b}$ is called a* **basic feasible solution**.

We are now ready to state the following fundamental theorem in the theory of linear programming:

THEOREM 21.3 *A vector $\mathbf{x}$ is an extreme point of the feasible set of a linear programming problem if and only if it is a basic feasible solution of the problem.*

This theorem will yield an algebraic technique for finding extreme points. The proof of this theorem is too long to present here. Instead, we refer the reader to any standard text in linear programming theory, such as S.I. Gass, *Linear Programming*, 4th ed., New York: McGraw-Hill Book Company, 1975.

EXAMPLE 21.5 Find an optimal solution of the linear programming problem:

$$\text{Maximize} \quad z = 2x_1 + 3x_2 - x_3 + 4x_5 + x_6$$

subject to

$$2x_1 \quad + x_3 + 4x_4 + 2x_5 = 20$$

$$x_1 + x_2 - x_3 + x_4 + 3x_5 = 10$$

and

$$x_1,\, x_2,\, x_3,\, x_4,\, x_5 \geq 0.$$

SOLUTION As we ask the reader to show in Exercise 21.9, the feasible set of this problem is bounded. Consequently, from Theorem 21.2 the objective function attains its maximum value at an extreme point. The linear system $Ax = b$ in the constraints is the same system considered in Example 21.4. Equations (21.10) give the ten basic solutions of this linear system. Of these ten, the eight listed in (21.11) are basic feasible solutions since they satisfy the nonnegativity condition. Thus, from Theorem 21.3, the extreme points of the feasible set are also given by Eq. (21.11). (Notice that these eight basic feasible solutions determine only five distinct extreme points.) Below each of the eight basic feasible solutions we have given the corresponding value of the objective function. From this we see that $z = 70$ is the maximum value of the objective function and an optimal solution is $x_1 = 0,\ x_2 = 30,\ x_3 = 20,\ x_4 = 0,\ x_5 = 0$.

$$
x_1 = \begin{bmatrix} \frac{10}{0} \\ 0 \\ 0 \\ 0 \end{bmatrix} \quad
x_2 = \begin{bmatrix} \frac{10}{0} \\ 0 \\ 0 \\ 0 \end{bmatrix} \quad
x_3 = \begin{bmatrix} \frac{10}{0} \\ 0 \\ \frac{0}{0} \\ 0 \end{bmatrix} \quad
x_4 = \begin{bmatrix} \frac{10}{0} \\ 0 \\ 0 \\ \frac{0}{0} \\ 0 \end{bmatrix} \quad
x_5 = \begin{bmatrix} 0 \\ 30 \\ 20 \\ 0 \\ 0 \end{bmatrix} \quad
x_6 = \begin{bmatrix} 0 \\ 5 \\ 0 \\ \frac{5}{0} \end{bmatrix} \quad
x_7 = \begin{bmatrix} 0 \\ 0 \\ 8 \\ 0 \\ 6 \end{bmatrix} \quad
x_8 = \begin{bmatrix} 0 \\ 0 \\ 0 \\ 4 \\ 2 \end{bmatrix}
$$

$$z = 20 \quad z = 20 \quad z = 20 \quad z = 20 \quad z = 70 \quad z = 35 \quad z = -2 \quad z = 18$$

(21.11)

The technique used in Example 21.5 can, in principle, be used to solve any linear programming problem. However, the number of basic feasible solutions quickly increases as the number of variables increases. For example, a linear programming problem with forty variables and twenty equality constraints could have over 130 billion basic feasible solutions. It would be completely impractical to find all of them, even with the fastest computer. Chapter 21 describes a practical alternative to this technique called the *simplex method*. We finish this chapter with a brief qualitative description of the simplex method to prepare for that chapter.

INTRODUCTION TO THE SIMPLEX METHOD

Let us introduce the following definition:

> **DEFINITION 21.6** *In a linear programming problem, two basic feasible solutions having m basic variables are said to be* **adjacent** *if they have m - 1 basic variables in common.*

Geometrically, extreme points corresponding to adjacent basic feasible solutions are connected by some "edge" of the feasible set. However, we shall not pursue this geometric interpretation.

EXAMPLE 21.5 (REVISITED)

Let us find the adjacent basic feasible solutions of the linear programming problem posed in Example 21.5. The eight basic feasible solutions of this problem are given in Eq. (21.11). In Fig. 21.5 we have drawn a graph of the eight basic feasible solutions. In this graph we have linked adjacent basic feasible solutions with a single line. For example, x_5 and x_7 are adjacent because they have $m - 1 = 2 - 1 = 1$ basic variable in common, namely x_3. On the other hand, x_4 and x_5 are not adjacent since they do not have $m - 1 = 1$ basic variable in common.

◆—◆—◆

The simplex method is a way of proceeding from one basic feasible solution to an adjacent basic feasible solution in such a way that the value of the objective function never decreases. This usually leads to a basic feasible solution for which the value of the objective function is as large as possible. We say "usually" because there is a slight complication caused by degeneracy which we shall describe below.

In Fig. 21.6(a) we have redrawn Fig. 21.5 and labeled each of the eight basic feasible solutions with the corresponding value of the objective function as given by (21.11). If somehow we generate x_8 as a basic feasible solution, one can show that the simplex

Figure 21.5

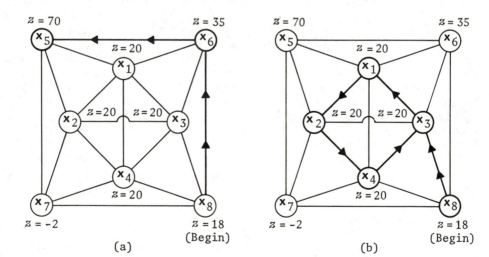

Figure 21.6

method will generate for us the adjacent basic feasible solution x_6. The value of the objective function is thereby increased from $z = 18$ to $z = 35$. From x_6, the simplex method will then generate the adjacent basic feasible solution x_5, where $z = 70$. At all basic feasible solutions adjacent to x_5, the value of the objective function is less than 70. Thus $z = 70$ is the maximum value of the objective function and x_5 is an optimal solution to the problem.

If any of the basic feasible solutions is degenerate, however, it is possible that the simplex method will not lead to an optimal solution. Since the simplex method only guarantees that the value of the objective function will not decrease, it is possible that the situation described in Figure 21.6(b) might arise. Here, we move around the basic feasible solutions x_3, x_1, x_2, and x_4 indefinitely without ever reaching the optimal basic feasible solution x_5. This phenomenon is known as *cycling*. Fortunately, it is not a serious problem in realistic linear programming problems. The circumstances which may produce cycling are rarely encountered in practice; and round-off error in a computer tends to destroy degeneracy so that a loop such as in Fig. 21.6(b) is eventually exited. At any rate, there are algorithms available which can eliminate cycling if it is suspected to be encountered in some specific problem.

EXERCISES

21.1 Convert the following linear programming problem to one in standard form:

$$\text{Minimize} \quad z = 2x_1 + 5x_2$$

subject to

$$3x_1 - 6x_2 \leq 2$$
$$x_1 + x_2 \leq 3$$
$$x_1 \geq 6$$
$$x_2 \leq 5$$

and

$$x_1, x_2 \geq 0.$$

21.2 Convert the following linear programming problem to one in standard form:

$$\text{Maximize} \quad z = -3x_1 + x_2 + x_3$$

subject to

$$3x_1 - 5x_2 + x_3 = 3$$
$$-2x_1 - x_2 \leq 2$$
$$x_1 \geq 5$$
$$x_2 \geq 2$$

and

$$x_1, x_2, x_3 \geq 0.$$

21.3 For the following linear programming problem in standard form:

$$\text{Maximize} \quad z = 2x_1 - x_2 + x_3 + 3x_4$$

subject to

$$x_1 + x_2 + 2x_3 = 2$$
$$2x_1 + 4x_3 + x_4 = 1$$

and

$$x_1, x_2, x_3, x_4 \geq 0,$$

(a) show that the feasible set is bounded,
(b) find all basic solutions of the linear system $Ax = b$,
(c) find all basic feasible solutions of the problem,
(d) find an optimal solution and the maximum value of the objective function,
(e) draw a graph of the basic feasible solutions, as in Fig. 21.5, in which adjacent basic feasible solutions are linked with a single line.

21.4 For the following linear programming problem in standard form:

$$\text{Maximize} \quad z = 2x_1 - 6x_2 + 3x_3$$

subject to

$$x_1 + 5x_2 + 3x_3 = 2$$
$$-x_1 + 2x_2 + 4x_3 = 3$$

and

$$x_1, x_2, x_3 \geq 0,$$

(a) find all basic solutions of the linear system $Ax = b$,
(b) show that the problem has no basic feasible solutions. (From this it follows that the feasible set of this problem is empty.)

21.5 For the following linear programming problem:

$$\text{Maximize} \quad z = 3x_1 + 2x_2 - x_3$$

subject to

$$2x_1 + 3x_2 + x_3 \leq 4$$

$$x_1 + 2x_2 + 3x_3 \leq 5$$

and

$$x_1, x_2, x_3 \geq 0,$$

(a) convert the problem to one in standard form,
(b) find all basic solutions of the linear system $Ax = b$ in the standard problem,
(c) find all basic feasible solutions of the standard problem,
(d) find an optimal solution of the standard problem and the maximum value of the objective function,
(e) draw a graph of the basic feasible solutions of the standard problem, as in Fig. 21.5, in which adjacent basic feasible solutions are linked with a single line.
(f) find an optimal solution of the original problem.

21.6 Repeat the instructions in Exercise 21.5 for the linear programming problem:

$$\text{Minimize} \quad z = 2x_1 - 3x_2 + x_3$$

subject to

$$x_1 - 2x_2 + 3x_3 \leq 5$$

$$2x_1 + x_2 - 2x_3 = 2$$

and

$$x_1, x_2, x_3 \geq 0.$$

21.7 Prove Theorem 21.1 as follows:
(a) Show that if $Ax_1 = b$ and $Ax_2 = b$, then $Ax = b$ if $x = tx_1 + (1 - t)x_2$ for any t in the interval $[0, 1]$.
(b) Show that if $x_1 \geq 0$ and $x_2 \geq 0$, then $x \geq 0$ if $x = tx_1 + (1 - t)x_2$ for any t in the interval $[0, 1]$.

21.8 Show that the feasible set of the linear programming problem in Example 21.2 is the set illustrated in Fig. 21.3.

21.9 Show that the feasible set of the linear programming problem in Example 21.5 is bounded. Proceed as follows:

(a) From the constraint

$$2x_1 + x_3 + 4x_4 + 2x_5 = 20$$

and the nonnegativity conditions $x_i \geq 0$ ($i = 1,2,3,4,5$) conclude that $x_i \leq 20$ for $i = 1,3,4,5$

(b) Add the two constraints together and conclude that $x_2 \leq 30$.

(c) From (a) and (b), conclude that $\|x\| \leq r$ for some positive number r.

21.10 Show that the linear system (21.7) can be written in the vector form (21.8).

Linear Programming 3: The Simplex Method

The simplex method is presented and applied to the solution of a certain class of linear programming problems.

PREREQUISITES: Gaussian elimination
                     Chapter 21: Linear Programming 2

INTRODUCTION

At the end of Chapter 21, we described the simplex method as a procedure for moving from one basic feasible solution of a linear programming problem to an adjacent basic feasible solution in such a way that the value of the objective function never decreases. In this chapter we shall describe the algebraic details of this procedure.

The general linear programming problem in n variables was stated as Problem 21.1 on page 304. However, in order to simplify the presentation of the simplex method, we shall restrict ourselves to linear programming problems having the following special form:

PROBLEM 22.1 *Find values of* $x_1, x_2, \ldots, x_n$ *which maximize*

$$z = c_1 x_1 + c_2 x_2 + \cdots + c_n x_n$$

subject to

$$a_{11} x_1 + a_{12} x_2 + \cdots + a_{1n} x_n \leq b_1$$

$$a_{21} x_1 + a_{22} x_2 + \cdots + a_{2n} x_n \leq b_2$$

$$\vdots$$

$$a_{m1} x_1 + a_{m2} x_2 + \cdots + a_{mn} x_n \leq b_m$$

and

$$x_i \geq 0, \qquad \text{for } i = 1, 2, \ldots, n,$$

where

$$b_j \geq 0, \qquad \text{for } j = 1, 2, \ldots, m.$$

In Problem 22.1, the condition that each of the m constraints be a $\leq$ inequality is not restrictive since it can easily be shown that any linear programming problem can always be written with all $\leq$ constraints. It is the condition $b_j \geq 0$ for $j = 1, 2, \ldots, m$ that is the real restriction. Nevertheless, a large class of practical problems are of this form, and the procedures developed in this chapter for Problem 22.1 are used in the application of the simplex method to the general linear programming problem.

To convert Problem 22.1 to one in standard form (see Problem 21.2 on page 304) we introduce m slack variables $x_{n+1}, x_{n+2}, \ldots, x_{n+m}$, one for each of the m constraints, to obtain

---

PROBLEM 22.2 *Find values of* $x_1, x_2, \ldots, x_{n+m}$ *which maximize*

$$z = c_1 x_1 + c_2 x_2 + \cdots + c_n x_n + 0 x_{n+1} + \cdots + 0 x_{n+m}$$

subject to

$$a_{11}x_1 + a_{12}x_2 + \cdots + a_{1n}x_n + x_{n+1} \qquad\qquad = b_1$$

$$a_{21}x_1 + a_{22}x_2 + \cdots + a_{2n}x_n \qquad\quad + x_{n+2} \qquad = b_2$$

$$\vdots \qquad\qquad\qquad\qquad\qquad \vdots$$

$$a_{m1}x_1 + a_{m2}x_2 + \cdots + a_{mn}x_n \qquad\qquad\quad + x_{n+m} = b_m$$

and

$$x_i \geq 0, \qquad \text{for } i = 1, 2, \ldots, n+m.$$

If we can find an optimal solution to Problem 22.2, then the values of the variables $x_1, x_2, \ldots, x_n$ will provide an optimal solution to problem 22.1.

THE SIMPLEX TABLEAU

In order to more clearly describe the steps in the simplex method, let us examine the following specific problem of the type we are considering in this chapter:

PROBLEM 22.3 *Find values of x_1, x_2, and x_3 which maximize*

$$z = 3x_1 - x_2 + 4x_3$$

subject to

$$2x_1 - x_2 + 3x_3 \leq 5$$

$$x_1 + 4x_2 - 2x_3 \leq 1$$

$$3x_1 \qquad + 6x_3 \leq 4$$

and

$$x_1, x_2, x_3 \geq 0.$$

To convert Problem 22.3 to one in standard form, we add slack variables x_4, x_5, x_6 to obtain

PROBLEM 22.4 *Find values of* x_1, x_2, x_3, x_4, x_5, *and* x_6 *which maximize*

$$z = 3x_1 - x_2 + 4x_3 + 0x_4 + 0x_5 + 0x_6$$

subject to

$$2x_1 - x_2 + 3x_3 + x_4 = 5$$
$$x_1 + 4x_2 - 2x_3 + x_5 = 1$$
$$3x_1 + 6x_3 + x_6 = 4$$

and

$$x_1, \ x_2, \ x_3, \ x_4, \ x_5, \ x_6 \geq 0.$$

For our purposes, it will be convenient to rephrase Problem 22.4 in the following equivalent form:

PROBLEM 22.5 *Find values of* x_1, x_2, x_3, x_4, x_5, x_6, *and* z *which satisfy*

$$2x_1 - x_2 + 3x_3 + x_4 = 5$$
$$x_1 + 4x_2 - 2x_3 + x_5 = 1$$
$$3x_1 + 6x_3 + x_6 = 4 \tag{22.1}$$
$$-3x_1 + x_2 - 4x_3 + z = 0$$

and such that x_1, x_2, x_3, x_4, x_5, x_6 *are nonnegative and* z *is as large as possible.*

In this formulation of the problem, z is treated as a variable on a par with x_1 through x_6, and the equation defining z in terms of the x_i is treated as an additional constraint. Thus our problem is to find a solution of the linear system of four equations in seven un-knowns given by (22.1) in which one of the variables, z, is as large as possible and the other six variables are nonnegative.

The usual procedure for solving a linear system of equations is to construct the augmented matrix of the system and apply Gaussian elimination or Gauss-Jordan elimination to it to put it in row-echelon form (or reduced row-echelon form). The row-echelon form then determines the augmented matrix of an equivalent linear system which is easily solved. The simplex method proceeds along similar lines. Namely, the augmented matrix of the linear system is constructed and a variation of Gaussian elimination, called *pivotal elimination*, is applied to obtain augmented matrices in which basic feasible solutions to the linear programming problem can be determined by inspection. Let us return to Problem 22.5 to see how this is done.

The augmented matrix of the linear system (22.1) is

$$
\begin{bmatrix}
2 & -1 & 3 & 1 & 0 & 0 & 0 & 5 \\
1 & 4 & -2 & 0 & 1 & 0 & 0 & 1 \\
3 & 0 & 6 & 0 & 0 & 1 & 0 & 4 \\
-3 & 1 & -4 & 0 & 0 & 0 & 1 & 0
\end{bmatrix}
\tag{22.2}
$$

From the way the 1's and 0's are distributed in the 4th, 5th, 6th, and 7th columns of this matrix, one particular solution of (22.1) can be seen by inspection; namely,

$$x_1 = 0, \ x_2 = 0, \ x_3 = 0, \ x_4 = 5, \ x_5 = 1, \ x_6 = 4, \ z = 0. \tag{22.3}$$

In terms of the corresponding linear programming problem, this is

$$
\mathbf{x}' =
\begin{bmatrix}
0 \\
0 \\
0 \\
5 \\
1 \\
4
\end{bmatrix}
\quad \text{and} \quad z = 0.
\tag{22.4}
$$

It is easily seen that $\mathbf{x}'$ is a basic feasible solution of linear programming Problem 22.4, according to Definition 21.5 on page 249. The three variables x_4, x_5, x_6 are the basic variables, and their corresponding values are found as the first three entries of the last column of (22.2). The value of z is the last entry of this column. That $\mathbf{x}'$ is a basic feasible solution is of crucial importance, since it is among the basic feasible solutions that we can hope to find an optimal solution. To see how we may go about finding another basic feasible solution, we rewrite (22.2) with some additional labeling:

Tableau 22.1

x_1	x_2	x_3	x_4	x_5	x_6	z		
2	-1	3	1	0	0	0	5	$= x_4$
1	4	-2	0	1	0	0	1	$= x_5$
3	0	6	0	0	1	0	4	$= x_6$
-3	1	-4	0	0	0	1	0	$= z$

Each column has been labeled with its corresponding variable from
the linear system (22.1). On the right, we have labeled the entries
of the last column with the corresponding variables whose values
they determine in the solution given by (22.3). We have also drawn
a vertical and horizontal line within the matrix in order to high-
light certain entries which will be useful to us later on.

In the field of linear programming, the augmented matrix is referred to as a *tableau*. In particular, Tableau 22.1 above is called
the *initial tableau* of the problem. We shall call the last row of
the tableau the *objective row*, since it arises from the objective
function of the original problem.

In Tableau 22.1 we have also shaded four particular columns.
It can be seen that these four columns are identical to the columns
of the 4×4 identity matrix. Indeed, it was exactly this fact that
permitted us to find the solution given in (22.3) so easily. We had
only to set those variables not corresponding to these four columns
equal to zero, and then the values of the variables corresponding to
the four columns were found in the last column of the augmented matrix.
This suggests a way of proceeding to a new solution of the linear
system. We apply appropriate elementary row operations to Tableau
22.1 to arrive at a new tableau which again contains the four columns of the 4×4 identity matrix, but this time in different position. To see how to do this, consider the following tableau, which
is just Tableau 22.1 with one of its entries shaded. (We postpone
for the moment a discussion of why this particular entry was chosen.)

Tableau 22.2

x_1	x_2	x_3	x_4	x_5	x_6	z		
2	-1	3	1	0	0	0	5	$= x_4$
1	4	-2	0	1	0	0	1	$= x_5$
3	0	6	0	0	1	0	4	$= x_6$
-3	1	-4	0	0	0	1	0	$= z$

Our objective will be to use elementary row operations to replace the shaded entry by a "1" and obtain zeros everywhere else in that column. To do this we first divide the third row by six to obtain:

Tableau 22.3

x_1	x_2	x_3	x_4	x_5	x_6	z	
2	-1	3	1	0	0	0	5
1	4	-2	0	1	0	0	1
1/2	0	1	0	0	1/6	0	2/3
-3	1	-4	0	0	0	1	0

Next we perform the following three elementary row operations:

1. Add -3 times the 3rd row to the 1st row.
2. Add 2 times the 3rd row to the 2nd row.
3. Add 4 times the 3rd row to the 4th row.

The result is the following tableau:

Tableau 22.4

x_1	x_2	x_3	x_4	x_5	x_6	z		
1/2	-1	0	1	0	-1/2	0	3	$= x_4$
2	4	0	0	1	1/3	0	7/3	$= x_5$
1/2	0	1	0	0	1/6	0	2/3	$= x_3$
-1	1	0	0	0	2/3	1	8/3	$= z$

It can be seen that Tableau 22.4 contains within it the four columns of the 4 × 4 identity matrix, though not in their usual order. Consequently, if we set the variables not associated with these columns equal to zero, we obtain the following solution to linear system (22.1):

$$x_1 = 0, \ x_2 = 0, \ x_3 = 2/3, \ x_4 = 3, \ x_5 = 7/3, \ x_6 = 0, \ z = 8/3. \quad (22.5)$$

As before, we have labeled the entries in the last column with the variables whose values they determine in this solution.

Solution (22.5) specifies the following vector solution of Problem 22.4:

$$\mathbf{x}'' = \begin{bmatrix} 0 \\ 0 \\ 2/3 \\ 3 \\ 7/3 \\ 0 \end{bmatrix} \quad \text{and} \quad z = 8/3. \tag{22.6}$$

We see that $\mathbf{x}''$ is a basic feasible solution with basic variables x_3, x_4, x_5. We also see that $\mathbf{x}'$ as given in (22.4) and $\mathbf{x}''$ are *adjacent basic feasible solutions* since they have $m - 1 = 2$ basic variables in common; namely, x_4 and x_5. That is, the elementary row operations we performed on Tableau 22.1 to obtain Tableau 22.4 took us from the basic feasible solution $\mathbf{x}'$ to the adjacent basic feasible solution $\mathbf{x}''$. In addition, the value of the objective function increased from $z = 0$ to $z = 8/3$.

The shaded entry in Tableau 22.2 is called the *pivot entry* of that particular tableau. The column in which it lies is called the *pivot column*, and the row in which it lies is called the *pivot row*. After the elementary row operations were applied to obtain Tableau 22.4, the pivot column contained all zeros, except for a one in the pivot entry position. Thus the pivot column was converted to one of the columns of the 4×4 identity matrix, and the variable x_3 corresponding to the pivot column was converted from a nonbasic variable to a basic variable. At the same time, the variable x_6 which labeled the pivot row, was converted from a basic variable to a nonbasic variable. That is, x_3 replaced x_6 as a basic variable in going from Tableau 22.1 to Tableau 22.4. For this reason, x_3 is called the *entering variable* and x_6 is called the *departing variable* of Tableau 22.1.

Let us see if we can increase the value of the objective function above the value $z = 8/3$ attained in Tableau 22.4. Below we have rewritten Tableau 22.4 with a pivot entry shaded. Again, we postpone until the next section a discussion of how this pivot entry was chosen.

Tableau 22.5

	x_1	x_2	x_3	x_4	x_5	x_6	z		
Depart	1/2	-1	0	1	0	-1/2	0	3	$= x_4$
x_5 ←	2	4	0	0	1	1/3	0	7/3	$= x_5$
	1/2	0	1	0	0	1/6	0	2/3	$= x_3$
	-1	1	0	0	0	2/3	1	8/3	z

Enter x_1

We apply the elementary row operations to the first column to convert it to a column with a "1" in the pivot position and zeros everywhere else. Thus, x_1 will be the entering variable and x_5 will be the departing variable, as the arrows in the tableau indicate. The reader can easily verify that the elementary row operations necessary will produce the next tableau:

Tableau 22.6

x_1	x_2	x_3	x_4	x_5	x_6	z		
0	-2	0	1	-1/4	-7/12	0	29/12	$= x_4$
1	2	0	0	1/2	1/6	0	7/6	$= x_1$
0	-1	1	0	-1/4	1/12	0	1/12	$= x_3$
0	3	0	0	1/2	5/6	1	23/6	$= z$

Setting those variables not corresponding to the shaded columns equal to zero yields the solution

$$x_1 = 7/6, \ x_2 = 0, \ x_3 = 1/12, \ x_4 = 29/12, \ x_5 = 0, \ x_6 = 0, \ z = 23/6.$$
$$(22.7)$$

For the linear programming problem, we then have the following basic feasible solution and objective function value:

$$\mathbf{x'''} = \begin{bmatrix} 7/6 \\ 0 \\ 1/12 \\ 29/12 \\ 0 \\ 0 \end{bmatrix} \quad \text{and} \quad z = 23/6. \quad (22.8)$$

As we shall show in the next section, the value $z = 23/6$ is the largest value the objective function can assume over the feasible set. Thus we have reached an optimal solution. For the solution to Problem 22.3, we discard the slack variables x_4, x_5, x_6 and write

$$x_1 = 7/6, \quad x_2 = 0, \quad x_3 = 1/12 \quad (22.9)$$

as the optimal solution, with the corresponding maximum $z = 23/6$ for the objective function.

This example illustrates the kinds of calculations required to implement the simplex method. In the next section we shall discuss how to choose the pivot entry and how to determine if an optimal solution has been reached.

STEPS IN THE SIMPLEX METHOD

In this section we shall outline the steps in the simplex method and give an example. In the next section we shall discuss their mathematical justification. The simplex method consists of the following five steps:

STEP 1 *Construct the initial tableau.*

STEP 2 *Test for optimality. If the tableau yields an optimal solution, then stop; otherwise, continue to Step 3.*

STEP 3 *Determine the pivot column.*

STEP 4 *Determine the pivot row.*

STEP 5 *Apply the elementary row operations to obtain all zeros in the pivot column, except for a "one" in the pivot row. Return to Step 2.*

The details for Steps 2, 3, and 4 are as follows:

Test for Optimality. If all of the entries in the objective row are nonnegative (ignoring the rightmost entry) the tableau determines an optimal solution.

Determination of Pivot Column. Choose the pivot column so that it contains the most negative entry of the objective row (ignoring the rightmost entry).

Determination of Pivot Row. Ignoring the objective row, divide each positive entry of the pivot column into the last entry in its row. Choose the pivot row to be one which yields the smallest such ratio.

The reader should return to the tableaus constructed in connection with Problem 22.5 to verify that the pivot entries in each tableau were selected according to the above rules, and also to verify that Tableau 22.6 determines an optimal solution to the problem.

Let us apply the simplex method as described above to the following example:

EXAMPLE 22.1 Find values of x_1, x_2, x_3 which maximize

$$z = 3x_1 + 4x_2 + 2x_3$$

subject to

$$3x_1 + 2x_2 + 4x_3 \leq 15$$

$$x_1 + 2x_2 + 3x_3 \leq 7$$

$$2x_1 + x_2 + x_3 \leq 6$$

and

$$x_1, x_2, x_3 \geq 0.$$

SOLUTION In standard form, this problem is

$$\text{Maximize} \quad z = 3x_1 + 4x_2 + 2x_3 + 0x_4 + 0x_5 + 0x_6$$

subject to

$$3x_1 + 2x_2 + 4x_3 + x_4 \qquad\qquad = 15$$

$$x_1 + 2x_2 + 3x_3 \qquad + x_5 \qquad = 7$$

$$2x_1 + x_2 + x_3 \qquad\qquad + x_6 = 6$$

and

$$x_1, x_2, x_3, x_4, x_5, x_6 \geq 0.$$

The initial tableau for the problem is then

Tableau 22.7

x_1	x_2	x_3	x_4	x_5	x_6	z		
3	2	4	1	0	0	0	15	$= x_4$
1	2	3	0	1	0	0	7	$= x_5$
2	1	1	0	0	1	0	6	$= x_6$
-3	-4	-2	0	0	0	1	0	$= z$

The objective row contains negative entries, so that the initial tableau does not determine an optimal solution. The most negative entry, -4, lies in the second column, so that the second column will be the pivot column. To determine the pivot row, we evaluate the following ratios:

1st row: 15/2 = 7½.   2nd row: 7/2 = 3½.   3rd row: 6/1 = 6.

The 2nd row yields the smallest ratio so that it will be the pivot row. So far we have the following:

Tableau 22.8

	x_1	x_2	x_3	x_4	x_5	x_6	z		
Depart	3	2	4	1	0	0	0	15	$= x_4$
x_5 ←	1	2	3	0	1	0	0	7	$= x_5$
	2	1	1	0	0	1	0	6	$= x_6$
	-3	-4	-2	0	0	0	1	0	$= z$

Enter x_2

We now apply the following elementary row operations to Tableau 22.8:

1. Divide the 2nd row by 2.
2. Add -2 times the 2nd row to the 1st row.
3. Add -1 times the 2nd row to the 3rd row.
4. Add 4 times the 2nd row to the 4th row.

The resulting tableau is the following:

Tableau 22.9

x_1	x_2	x_3	x_4	x_5	x_6	z		
2	0	1	1	-1	0	0	8	$= x_4$
1/2	1	3/2	0	1/2	0	0	7/2	$= x_2$
3/2	0	-1/2	0	-1/2	1	0	5/2	$= x_6$
-1	0	4	0	2	0	1	14	$= z$

The objective row still contains a negative entry, so that we have not yet reached an optimal solution. The new pivot column is the first since it contains the only negative entry of the objective row. To determine the pivot row, we evaluate the following ratios:

1st row: 8/2 = 4. 2nd row: (7/2)/(1/2) = 7. 3rd row: (5/2)/(3/2) = $1\frac{2}{3}$.

The third row yields the smallest ratio, so that it is the new pivot row. So far, we have

Tableau 22.10

	x_1	x_2	x_3	x_4	x_5	x_6	z		
	2	0	1	1	-1	0	0	8	$= x_4$
Depart	1/2	1	3/2	0	1/2	0	0	7/2	$= x_2$
x_6 ←	3/2	0	-1/2	0	-1/2	1	0	5/2	$= x_6$
	-1	0	4	0	2	0	1	14	$= z$

Enter x_1

We now apply the following elementary row operations to Tableau 22.10:

1. Divide the 3rd row by 3/2.
2. Add -2 times the 3rd row to the 1st row.
3. Add -½ times the 3rd row to the 2nd row.
4. Add 1 times the 3rd row to the 4th row.

The resulting tableau is

Tableau 22.11

x_1	x_2	x_3	x_4	x_5	x_6	z		
0	0	5/3	1	-1/3	-4/3	0	14/3	$= x_4$
0	1	5/3	0	2/3	-1/3	0	8/3	$= x_2$
1	0	-1/3	0	-1/3	2/3	0	5/3	$= x_1$
0	0	11/3	0	5/3	2/3	1	47/3	$= z$

The objective row of this tableau does not contain any negative entries and so this tableau determines an optimal solution. The basic variables in the optimal basic feasible solution are x_1, x_2, and x_4 as the righthand labeling indicates. The optimal solution is

$$x_1 = 5/3, \quad x_2 = 8/3, \quad x_3 = 0, \quad x_4 = 14/3, \quad x_5 = 0, \quad x_6 = 0, \quad z = 47/3.$$

For the original problem posed in this example, we discard the slack variables x_4, x_5, x_6 and simply write

$$x_1 = 5/3, \quad x_2 = 8/3, \quad x_3 = 0, \quad z = 47/3.$$

We conclude this section with some remarks concerning complications which may arise in the use of the simplex method as we have described it:

1. In Step 3 it is possible that there is a tie for the most negative entry in the objective row. In that case, any one of them may be chosen, and no complications arise.

2. In Step 4 it is possible that there is more than one row with the same smallest ratio. In that case, any one of them may be used to determine the pivot row and no complications arise in the calculations. However, if such a tie arises, it can be shown that the basic feasible solution determined by the next tableau will be degenerate (i.e., will have a basic variable whose value is zero). As discussed in the last chapter, it is degeneracy which may bring about cycling. But as we also mentioned, it is more a theoretical problem than a practical problem.

3. In Step 4 it is possible that no entry in the pivot column is positive, in which case our technique for evaluating the pivot row is meaningless. It can be shown that if this situation arises, the problem has an *unbounded solution*.

JUSTIFICATION OF THE STEPS IN THE SIMPLEX METHOD

Let us return to the linear programming problem in $n + m$ variables posed in Problem 22.2. Suppose at some point in our calculations we have arrived at Tableau 22.12. (In Exercise 22.11, we ask the reader to show that in any tableau the column labeled "z" always has the form indicated.) Thus the current basic variables are x_{B1}, x_{B2}, ..., x_{Bm} with corresponding values $d_1, d_2, \ldots, d_m$, and the current value of the objective variable is w. Let us see if the entry y_{rs} would make a suitable pivot entry. The entering variable would be x_s and the departing variable would be x_{Br}. If $y_{rs} \neq 0$, the elementary row operations in Step 5 of the simplex method will produce a tableau having the form of Tableau 22.13.

Tableau 22.12

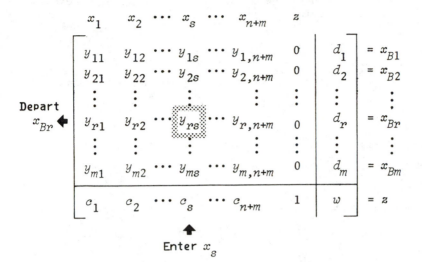

Enter x_s

Tableau 22.13

$$
\begin{bmatrix}
y_{11}^* & y_{12}^* & \cdots & 0 & \cdots & y_{1,n+m}^* & 0 & \vline & d_1 - y_{1s}d_r/y_{rs} \\
y_{21}^* & y_{22}^* & \cdots & 0 & \cdots & y_{2,n+m}^* & 0 & \vline & d_2 - y_{2s}d_r/y_{rs} \\
\vdots & \vdots & & \vdots & & \vdots & \vdots & \vline & \vdots \\
y_{r1}^* & y_{r2}^* & \cdots & 1 & \cdots & y_{r,n+m}^* & 0 & \vline & d_r/y_{rs} \\
\vdots & \vdots & & \vdots & & \vdots & \vdots & \vline & \vdots \\
y_{m1}^* & y_{m2}^* & \cdots & 0 & \cdots & y_{m,n+m}^* & 0 & \vline & d_m - y_{ms}d_r/y_{rs} \\
\hline
c_1^* & c_2^* & \cdots & 0 & \cdots & c_{n+m}^* & 1 & \vline & w - c_s d_r/y_{rs}
\end{bmatrix}
\begin{matrix}
= x_{B1} \\
= x_{B2} \\
\vdots \\
= x_s \\
\vdots \\
= x_{Bm} \\
\\
= z
\end{matrix}
$$

The pivot column now contains all zeros except for the "1" in the previous pivot entry position. All of the other entries in the tableau have new values, which we indicate with asterisks, except in the last column where we have explicitly written the new values in terms of the entries in Tableau 22.12. First, let us examine the value of the entering variable x_s:

$$x_s = d_r/y_{rs} \,. \tag{22.10}$$

Since $d_r \geq 0$, we see that we must have $y_{rs} > 0$ in order to satisfy the constraint $x_s \geq 0$. Let us list this fact as

OBSERVATION 22.1 *The pivot entry must be positive in order that the new tableau determine a feasible solution.*

The remaining basic variables have values given by

$$x_{Bi} = d_i - y_{is}d_r/y_{rs} \qquad \text{for} \quad i = 1, 2, \ldots, m; \ i \neq s. \tag{22.11}$$

We must have $x_{Bi} \geq 0$ for the new solution to be feasible. Now if for some i we have $y_{is} \leq 0$, then (22.11) states that for that i, $x_{Bi} \geq 0$ since $d_i \geq 0$, $d_r \geq 0$, and $y_{rs} > 0$. On the other hand, if for some i we have $y_{is} > 0$, then (21.11) requires that

$$d_i - y_{is}d_r/y_{rs} \geq 0 \tag{22.12}$$

in order that $x_{Bi} \geq 0$ for that i. Equation (22.12) can also be written as

$$\frac{d_r}{y_{rs}} \leq \frac{d_i}{y_{is}} \,. \tag{22.13}$$

In other words, Eq. (22.13) must be satisfied for all those i for which $y_{is} > 0$ in order that the new tableau determine a feasible solution. We state this as

OBSERVATION 22.2 *In order that the new tableau determine a feasible solution, the following must be true: The ratio of the element in the rightmost column of the pivot row to the pivot entry must be the smallest of the corresponding ratios in all of the other rows which contain positive entries in the pivot column (ignoring the objective row).*

Next, let us examine the new value w^* of the objective function. From Tableau 22.13, we see that

$$w^* = w - c_s d_r / y_{rs} \,. \tag{22.14}$$

Ideally, we would want the increase in the objective function

$$w^* - w = -c_s d_r / y_{rs} \tag{22.15}$$

to be as large as possible. But this would require that we compute the quantities

$$-c_s d_r / y_{rs}$$

for all possible values of r and s to find the largest one. Usually this is not done because of the large number of calculations this would require. Instead, the entering variable x_s is chosen so that c_s is as negative as possible. Since $d_r \geq 0$ and $y_{rs} > 0$, Eq. (22.15) then guarantees that

$$w^* - w \geq 0 \,;$$

i.e., that the objective function does not decrease. As discussed in the last chapter, this always eventually leads to an optimal solution, except for the very remote possibility of cycling. Let us summarize this as follows:

OBSERVATION 22.3 *Choose the pivot column so that it contains the most negative entry of the objective row (ignoring the entry in the rightmost column).*

Equation (22.15) also tells us the following: If all of the c's are nonnegative then the value of the objective function cannot increase, regardless of the choice of pivot entry. In this case, we must have already attained the maximum value of the objective function. Thus, we have

OBSERVATION 22.4 *If all of the entries in the objective row, except for the rightmost entry, are nonnegative, the tableau determines an optimal solution.*

The above four observations justify the steps in the simplex method.

Readers interested in pursuing linear programming in more detail are referred to the following texts:

S. I. Gass, *Linear Programming*, 4th ed. New York: McGraw-Hill Book Company, 1975.

L. Cooper and D. Steinberg, *Methods and Applications of Linear Programming*, Philadelphia: W. B. Saunders Company, 1974.

EXERCISES

In Exercises 22.1 to 22.6 solve the given linear programming problem by the simplex method.

22.1   Maximize $z = 3x_1 + 4x_2$

subject to

$$2x_1 + 3x_2 \leq 7$$
$$5x_1 + 2x_2 \leq 3$$

and

$$x_1,\ x_2 \geq 0.$$

22.2   Maximize $z = 2x_1 + x_2$

subject to

$$3x_1 + 2x_2 \leq 4$$
$$3x_1 + x_2 \leq 3$$
$$2x_1 \qquad \leq 3$$

and

$$x_1,\ x_2 \geq 0.$$

22.3   Maximize $z = 3x_1 - 2x_2 + 6x_3$

subject to

$$2x_1 - 5x_2 + x_3 \leq 2$$
$$x_1 + x_2 + x_3 \leq 5$$

and

$$x_1,\ x_2,\ x_3 \geq 0.$$

22.4 $\qquad$ Maximize $\quad z = 2x_1 + x_2 - x_3$

subject to

$$2x_1 - 3x_2 + x_3 \leq 2$$
$$x_1 + 5x_2 - 2x_3 \leq 4$$
$$2x_1 - 4x_2 - x_3 \leq 3$$

and

$$x_1, \ x_2, \ x_3 \geq 0.$$

22.5 $\qquad$ Maximize $\quad z = 3x_1 - 2x_2 - x_3 + x_4$

subject to

$$2x_1 - 3x_2 + x_3 - x_4 \leq 6$$
$$x_1 + 2x_2 - x_3 + 2x_4 \leq 4$$

and

$$x_1, \ x_2, \ x_3, \ x_4 \geq 0$$

22.6 $\qquad$ Maximize $\quad z = x_1 - 2x_2 + 3x_3 + x_4$

subject to

$$x_1 - 2x_2 + x_3 + 3x_4 \leq 8$$
$$2x_1 + 3x_2 - x_3 + 2x_4 \leq 5$$
$$x_1 + x_2 - 3x_3 + 4x_4 \leq 6$$

and

$$x_1, \ x_2, \ x_3, \ x_4 \geq 0$$

22.7 Solve Example 20.1 by the simplex method.

22.8 Solve Exercise 20.1 by the simplex method.

22.9 Solve Exercise 20.6 by the simplex method.

22.10 Solve Exercise 20.7 by the simplex method.

22.11 Show that in any tableau the column labeled "z" always has the form indicated in Tableau 22.12.

Answers to Exercises

Chapter 1, page 8

1.1 (a) $y = 3x - 4$

(b) $y = -2x + 1$

1.2 (a) $x^2 + y^2 - 4x - 6y + 4 = 0$ or $(x - 2)^2 + (y - 3)^2 = 9$

(b) $x^2 + y^2 + 2x - 4y - 20 = 0$ or $(x + 1)^2 + (y - 2)^2 = 25$

1.3 $4x^2 + 8xy + 4y^2 - 8x + 4y = 0$ (a parabola)

1.4 (a) $x + 2y + z = 0$

(b) $-x + y - 2z + 1 = 0$

1.5 (a) $x^2 + y^2 + z^2 - 2x - 4y - 2z = -2$ or $(x-1)^2 + (y-2)^2 + (z-1)^2 = 4$

(b) $x^2 + y^2 + z^2 - 2x - 2y = 3$ or $(x - 1)^2 + (y - 1)^2 + z^2 = 5$

1.9
$$\begin{vmatrix} y & x^2 & x & 1 \\ y_1 & x_1^2 & x_1 & 1 \\ y_2 & x_2^2 & x_2 & 1 \\ y_3 & x_3^3 & x_3 & 1 \end{vmatrix} = 0$$

Chapter 3, page 33

3.1

3.2

3.3

3.4

3.5

3.6

Chapter 4, page 50

4.1 (a) $\begin{bmatrix} 0 & 0 & 0 & 1 \\ 1 & 0 & 1 & 1 \\ 1 & 1 & 0 & 1 \\ 0 & 0 & 0 & 0 \end{bmatrix}$ (b) $\begin{bmatrix} 0 & 1 & 1 & 0 & 0 \\ 0 & 0 & 0 & 0 & 1 \\ 1 & 0 & 0 & 1 & 0 \\ 0 & 0 & 1 & 0 & 0 \\ 0 & 0 & 1 & 0 & 0 \end{bmatrix}$ (c) $\begin{bmatrix} 0 & 1 & 0 & 1 & 0 & 0 \\ 1 & 0 & 0 & 0 & 0 & 0 \\ 0 & 1 & 0 & 1 & 1 & 1 \\ 0 & 0 & 0 & 0 & 0 & 1 \\ 0 & 0 & 0 & 0 & 0 & 1 \\ 0 & 0 & 1 & 0 & 1 & 0 \end{bmatrix}$

4.2 (a) (b)

(c)

4.3 (a)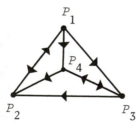

(b) 1-step: $P_1 \to P_2$

2-step: $P_1 \to P_4 \to P_2$
$P_1 \to P_3 \to P_2$

3-step: $P_1 \to P_2 \to P_1 \to P_2$
$P_1 \to P_3 \to P_4 \to P_2$
$P_1 \to P_4 \to P_3 \to P_2$

(c) 1-step: $P_1 \to P_4$

2-step: $P_1 \to P_3 \to P_4$

3-step: $P_1 \to P_2 \to P_1 \to P_4$
$P_1 \to P_4 \to P_3 \to P_4$

4.4 (a) $\{P_1, P_2, P_3\}$ (b) $\{P_3, P_4, P_5\}$ (c) $\{P_2, P_4, P_6, P_8\}$ and $\{P_4, P_5, P_6\}$

4.5 (a) None (b) $\{P_3, P_4, P_6\}$

4.6
$$\begin{bmatrix} 0 & 0 & 1 & 1 \\ 1 & 0 & 0 & 0 \\ 0 & 1 & 0 & 1 \\ 0 & 1 & 0 & 0 \end{bmatrix}$$
Power of P_1 = 5
Power of P_2 = 3
Power of P_3 = 4
Power of P_4 = 2

4.7 First, A; second, B and E (tie); fourth, C; fifth, D.

Chapter 5, page 65

5.1 (a) $-5/8$

(b) $\begin{bmatrix} 0 & 1 & 0 \end{bmatrix}$

(c) $\begin{bmatrix} 1 & 0 & 0 & 0 \end{bmatrix}^t$

5.2 Let $A = \begin{bmatrix} 1 & 1 \\ 1 & 1 \end{bmatrix}$ for example.

5.3 (a) $\mathbf{p}^* = \begin{bmatrix} 0 & 1 \end{bmatrix}$, $\mathbf{q}^* = \begin{bmatrix} 0 \\ 1 \end{bmatrix}$, $v = 3$

(b) $\mathbf{p}^* = \begin{bmatrix} 0 & 1 & 0 \end{bmatrix}$, $\mathbf{q}^* = \begin{bmatrix} 1 \\ 0 \end{bmatrix}$, $v = 2$

(c) $\mathbf{p}^* = \begin{bmatrix} 0 & 0 & 1 \end{bmatrix}$, $\mathbf{q}^* = \begin{bmatrix} 0 \\ 1 \\ 0 \end{bmatrix}$, $v = 2$

(d) $\mathbf{p}^* = \begin{bmatrix} 0 & 1 & 0 & 0 \end{bmatrix}$, $\mathbf{q}^* = \begin{bmatrix} 1 \\ 0 \\ 0 \end{bmatrix}$, $v = -2$

5.4 (a) $\mathbf{p}^* = \begin{bmatrix} 5/8 & 3/8 \end{bmatrix}$, $\mathbf{q}^* = \begin{bmatrix} 1/8 \\ 7/8 \end{bmatrix}$, $v = 27/8$

(b) $\mathbf{p}^* = \begin{bmatrix} 2/3 & 1/3 \end{bmatrix}$, $\mathbf{q}^* = \begin{bmatrix} 1/6 \\ 5/6 \end{bmatrix}$, $v = 70/3$

(c) $\mathbf{p}^* = \begin{bmatrix} 1 & 0 \end{bmatrix}$, $\mathbf{q}^* = \begin{bmatrix} 1 \\ 0 \end{bmatrix}$, $v = 3$

(d) $p^* = \begin{bmatrix} 3/5 & 2/5 \end{bmatrix}$, $q^* = \begin{bmatrix} 3/5 \\ 2/5 \end{bmatrix}$, $v = 19/5$

(e) $p^* = \begin{bmatrix} 3/13 & 10/13 \end{bmatrix}$, $q^* = \begin{bmatrix} 1/13 \\ 12/13 \end{bmatrix}$, $v = -29/13$

5.5 $p^* = \begin{bmatrix} 13/20 & 7/20 \end{bmatrix}$, $q^* = \begin{bmatrix} 11/20 \\ 9/20 \end{bmatrix}$, $v = -3/20$

Chapter 6, page 78

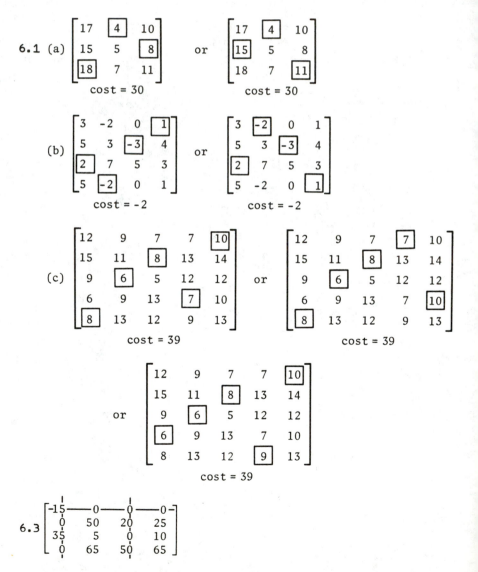

6.1 (a)

$$\begin{bmatrix} 17 & \boxed{4} & 10 \\ 15 & 5 & \boxed{8} \\ \boxed{18} & 7 & 11 \end{bmatrix} \quad \text{or} \quad \begin{bmatrix} 17 & \boxed{4} & 10 \\ \boxed{15} & 5 & 8 \\ 18 & 7 & \boxed{11} \end{bmatrix}$$

cost = 30   cost = 30

(b)

$$\begin{bmatrix} 3 & -2 & 0 & \boxed{1} \\ 5 & 3 & \boxed{-3} & 4 \\ \boxed{2} & 7 & 5 & 3 \\ 5 & \boxed{-2} & 0 & 1 \end{bmatrix} \quad \text{or} \quad \begin{bmatrix} 3 & \boxed{-2} & 0 & 1 \\ 5 & 3 & \boxed{-3} & 4 \\ \boxed{2} & 7 & 5 & 3 \\ 5 & -2 & 0 & \boxed{1} \end{bmatrix}$$

cost = -2   cost = -2

(c)

$$\begin{bmatrix} 12 & 9 & 7 & 7 & \boxed{10} \\ 15 & 11 & \boxed{8} & 13 & 14 \\ 9 & \boxed{6} & 5 & 12 & 12 \\ 6 & 9 & 13 & \boxed{7} & 10 \\ \boxed{8} & 13 & 12 & 9 & 13 \end{bmatrix} \quad \text{or} \quad \begin{bmatrix} 12 & 9 & 7 & \boxed{7} & 10 \\ 15 & 11 & \boxed{8} & 13 & 14 \\ 9 & \boxed{6} & 5 & 12 & 12 \\ 6 & 9 & 13 & 7 & \boxed{10} \\ \boxed{8} & 13 & 12 & 9 & 13 \end{bmatrix}$$

cost = 39   cost = 39

or

$$\begin{bmatrix} 12 & 9 & 7 & 7 & \boxed{10} \\ 15 & 11 & \boxed{8} & 13 & 14 \\ 9 & \boxed{6} & 5 & 12 & 12 \\ \boxed{6} & 9 & 13 & 7 & 10 \\ 8 & 13 & 12 & \boxed{9} & 13 \end{bmatrix}$$

cost = 39

6.3

$$\begin{bmatrix} -15 & 0 & 0 & 0 \\ 0 & 50 & 20 & 25 \\ 35 & 5 & 0 & 10 \\ 0 & 65 & 50 & 65 \end{bmatrix}$$

6.4

$$\begin{bmatrix} 0 & 3 & 0 & 4 & \boxed{0} \\ 4 & \boxed{0} & 6 & 1 & 5 \\ 3 & 1 & \boxed{0} & 4 & 0 \\ \boxed{0} & 1 & 6 & 2 & 1 \\ 0 & 0 & 0 & \boxed{0} & 3 \end{bmatrix}$$

6.5 (a) Bidder 1 receives Coin 3 ($210)
　　Bidder 2 receives Coin 1 ($175)
　　Bidder 3 receives Coin 2 ($ 85)
　　Bidder 4 receives nothing
　　Bidder 5 receives Coin 4 ($160)

　　Sum of bids = $630

(b) Repeat bidder 2's row twice in the cost matrix; add two dummy coins to end up with a 6×6 cost matrix. Optimal assignment is

　　Bidder 1 receives nothing
　　Bidder 2 receives Coins 1 and 3 ($175 + $230)
　　Bidder 3 receives Coin 2 ($85)
　　Bidder 4 receives nothing
　　Bidder 5 receives Coin 4 ($160)

　　Sum of bids = $650

6.6 P - Pete　　　2B - Jill or Ann　　LF - Herb
　　C - Liz　　　3B - Alex　　　　CF - John or Lois
　　1B - Sam　　SS - Ann or Jill　　RF - Lois or John

Chapter 7, page 93

7.1 (a) $x^{(1)} = \begin{bmatrix} .4 \\ .6 \end{bmatrix}$, $x^{(2)} = \begin{bmatrix} .46 \\ .54 \end{bmatrix}$, $x^{(3)} = \begin{bmatrix} .454 \\ .546 \end{bmatrix}$,

$x^{(4)} = \begin{bmatrix} .4546 \\ .5454 \end{bmatrix}$, $x^{(5)} = \begin{bmatrix} .45454 \\ .54546 \end{bmatrix}$

(b) P is regular since all entries of P are positive;

$q = \begin{bmatrix} 5/11 \\ 6/11 \end{bmatrix}$

7.2 (a) $x^{(1)} = \begin{bmatrix} .7 \\ .2 \\ .1 \end{bmatrix}$, $x^{(2)} = \begin{bmatrix} .23 \\ .52 \\ .25 \end{bmatrix}$, $x^{(3)} = \begin{bmatrix} .273 \\ .396 \\ .331 \end{bmatrix}$

(b) P is regular since all entries of P are positive;

$$q = \begin{bmatrix} 22/72 \\ 29/72 \\ 21/72 \end{bmatrix}$$

7.3 (a) $\begin{bmatrix} 9/17 \\ 8/17 \end{bmatrix}$ (b) $\begin{bmatrix} 26/45 \\ 19/45 \end{bmatrix}$ (c) $\begin{bmatrix} 3/19 \\ 4/19 \\ 12/19 \end{bmatrix}$

7.4 (a) $P^n = \begin{bmatrix} \left(\frac{1}{2}\right)^n & 0 \\ 1 - \left(\frac{1}{2}\right)^n & 1 \end{bmatrix}$, $n = 1, 2, \ldots$ by induction.

Thus no integer power of P has all positive entries.

(b) $P^n \to \begin{bmatrix} 0 & 0 \\ 1 & 1 \end{bmatrix}$ as n increases, and so $P^n x^{(0)} \to \begin{bmatrix} 0 \\ 1 \end{bmatrix}$ for any $x^{(0)}$ as n increases.

(c) The entries of the limiting vector $\begin{bmatrix} 0 \\ 1 \end{bmatrix}$ are not all positive.

7.6 $P^2 = \begin{bmatrix} \frac{1}{2} & \frac{1}{4} & \frac{1}{4} \\ \frac{1}{4} & \frac{1}{2} & \frac{1}{4} \\ \frac{1}{4} & \frac{1}{4} & \frac{1}{2} \end{bmatrix}$ has all positive entries; $q = \begin{bmatrix} 1/3 \\ 1/3 \\ 1/3 \end{bmatrix}$

7.7 10/13

7.8 $54\frac{1}{6}\%$ in region 1, $16\frac{2}{3}\%$ in region 2, and $29\frac{1}{6}\%$ in region 3.

Chapter 8, page 105

8.1 (a) $\begin{bmatrix} 2 \\ 3 \end{bmatrix}$, (b) $\begin{bmatrix} 6 \\ 5 \\ 6 \end{bmatrix}$, (c) $\begin{bmatrix} 78 \\ 54 \\ 79 \end{bmatrix}$.

8.2 (a) Use Corollary 8.1; all row sums are less than one.

(b) Use Corollary 8.2; all column sums are less than one.

(c) Use Theorem 8.3 with $x = \begin{bmatrix} 2 \\ 1 \\ 1 \end{bmatrix} > Cx = \begin{bmatrix} 1.9 \\ .9 \\ .9 \end{bmatrix}$.

8.3 E^2 has all positive entries.

8.4 Price of tomatoes: \$120.00.   Price of corn: \$100.00.
Price of lettuce: \$106.67.

8.5 \$1256 for the CE, \$1448 for the EE, \$1556 for the ME.

Chapter 9, page 120

9.1 The second class; \$15,000.

9.2 \$223

9.3 $1 : 1.90 : 3.02 : 4.24 : 5.00$

9.4 $s/(g_1^{-1} + g_2^{-1} + \cdots + g_{k-1}^{-1})$

9.5 $1 : 2 : 3 : \cdots : n-1$

Chapter 10, page 130

10.1 (a) $\begin{bmatrix} 0 & 1 & 1 & 0 \\ 0 & 0 & 1 & 1 \\ 0 & 0 & 0 & 0 \end{bmatrix}$

(b) $\begin{bmatrix} 0 & 1\frac{1}{2} & 1\frac{1}{2} & 0 \\ 0 & 0 & \frac{1}{2} & \frac{1}{2} \\ 0 & 0 & 0 & 0 \end{bmatrix}$

(c) $\begin{bmatrix} -2 & -1 & -1 & -2 \\ -1 & -1 & 0 & 0 \\ 3 & 3 & 3 & 3 \end{bmatrix}$

(d) $\begin{bmatrix} 0 & .866 & 1.366 & .500 \\ 0 & -.500 & .366 & .866 \\ 0 & 0 & 0 & 0 \end{bmatrix}$

10.2 (b) $(0, 0, 0)$, $(1, 0, 0)$, $(1\frac{1}{2}, 1, 0)$, and $(\frac{1}{2}, 1, 0)$

(c) $(0, 0, 0)$, $(1, .6, 0)$

$(1, 1.6, 0)$, $(0, 1, 0)$

10.3 (a) $\begin{bmatrix} 1 & 0 & 0 \\ 0 & -1 & 0 \\ 0 & 0 & 1 \end{bmatrix}$

(b) $\begin{bmatrix} -1 & 0 & 0 \\ 0 & 1 & 0 \\ 0 & 0 & 1 \end{bmatrix}$

(c) $\begin{bmatrix} 1 & 0 & 0 \\ 0 & 1 & 0 \\ 0 & 0 & -1 \end{bmatrix}$

10.4 (a) $M_1 = \begin{bmatrix} 1/2 & 0 & 0 \\ 0 & 2 & 0 \\ 0 & 0 & 1/3 \end{bmatrix}$, $M_2 = \begin{bmatrix} \frac{1}{2} & \frac{1}{2} & \cdots & \frac{1}{2} \\ 0 & 0 & \cdots & 0 \\ 0 & 0 & \cdots & 0 \end{bmatrix}$,

$M_3 = \begin{bmatrix} 1 & 0 & 0 \\ 0 & \cos 20° & -\sin 20° \\ 0 & \sin 20° & \cos 20° \end{bmatrix}$, $M_4 = \begin{bmatrix} \cos(-45°) & 0 & \sin(-45°) \\ 0 & 1 & 0 \\ -\sin(-45°) & 0 & \cos(-45°) \end{bmatrix}$,

$M_5 = \begin{bmatrix} 0 & -1 & 0 \\ 1 & 0 & 0 \\ 0 & 0 & 1 \end{bmatrix}$

(b) $P' = M_5 M_4 M_3 (M_1 P + M_2)$

10.5 (a) $M_1 = \begin{bmatrix} .3 & 0 & 0 \\ 0 & .5 & 0 \\ 0 & 0 & 1 \end{bmatrix}$, $M_2 = \begin{bmatrix} 1 & 0 & 0 \\ 0 & \cos 45° & -\sin 45° \\ 0 & \sin 45° & \cos 45° \end{bmatrix}$,

$M_3 = \begin{bmatrix} 1 & 1 & \cdots & 1 \\ 0 & 0 & \cdots & 0 \\ 0 & 0 & \cdots & 0 \end{bmatrix}$, $M_4 = \begin{bmatrix} \cos 35° & 0 & \sin 35° \\ 0 & 1 & 0 \\ -\sin 35° & 0 & \cos 35° \end{bmatrix}$,

$M_5 = \begin{bmatrix} \cos(-45) & -\sin(-45) & 0 \\ \sin(-45) & \cos(-45) & 0 \\ 0 & 0 & 1 \end{bmatrix}$, $M_6 = \begin{bmatrix} 0 & 0 & \cdots & 0 \\ 0 & 0 & \cdots & 0 \\ 1 & 1 & \cdots & 1 \end{bmatrix}$,

$M_7 = \begin{bmatrix} 2 & 0 & 0 \\ 0 & 1 & 0 \\ 0 & 0 & 1 \end{bmatrix}$

(b) $P' = M_7 (M_5 M_4 (M_2 M_1 P + M_3) + M_6)$

10.6 $R_1 = \begin{bmatrix} \cos \beta & 0 & \sin \beta \\ 0 & 1 & 0 \\ -\sin \beta & 0 & \cos \beta \end{bmatrix}$, $R_2 = \begin{bmatrix} \cos \alpha & -\sin \alpha & 0 \\ \sin \alpha & \cos \alpha & 0 \\ 0 & 0 & 1 \end{bmatrix}$,

$R_3 = \begin{bmatrix} \cos \theta & 0 & \sin \theta \\ 0 & 1 & 0 \\ -\sin \theta & 0 & \cos \theta \end{bmatrix}$, $R_4 = \begin{bmatrix} \cos \alpha & \sin \alpha & 0 \\ -\sin \alpha & \cos \alpha & 0 \\ 0 & 0 & 1 \end{bmatrix}$,

$R_5 = \begin{bmatrix} \cos \beta & 0 & -\sin \beta \\ 0 & 1 & 0 \\ \sin \beta & 0 & \cos \beta \end{bmatrix}$

10.7 (a) $M = \begin{bmatrix} 1 & 0 & 0 & x_0 \\ 0 & 1 & 0 & y_0 \\ 0 & 0 & 1 & z_0 \\ 0 & 0 & 0 & 1 \end{bmatrix}$ (b) $\begin{bmatrix} 1 & 0 & 0 & -5 \\ 0 & 1 & 0 & 9 \\ 0 & 0 & 1 & -3 \\ 0 & 0 & 0 & 1 \end{bmatrix}$

Chapter 11, page 147

11.1 (a)
$$\begin{bmatrix} t_1 \\ t_2 \\ t_3 \\ t_4 \end{bmatrix} = \begin{bmatrix} 0 & \frac{1}{4} & \frac{1}{4} & 0 \\ \frac{1}{4} & 0 & 0 & \frac{1}{4} \\ \frac{1}{4} & 0 & 0 & \frac{1}{4} \\ 0 & \frac{1}{4} & \frac{1}{4} & 0 \end{bmatrix} \begin{bmatrix} t_1 \\ t_2 \\ t_3 \\ t_4 \end{bmatrix} + \begin{bmatrix} 0 \\ \frac{1}{2} \\ 0 \\ \frac{1}{2} \end{bmatrix}$$

(b) $\quad t = \begin{bmatrix} 1/4 \\ 3/4 \\ 1/4 \\ 3/4 \end{bmatrix}$

(c) $t^{(1)} = \begin{bmatrix} 0 \\ \frac{1}{2} \\ 0 \\ \frac{1}{2} \end{bmatrix}$, $\quad t^{(2)} = \begin{bmatrix} 1/8 \\ 5/8 \\ 1/8 \\ 5/8 \end{bmatrix}$, $\quad t^{(3)} = \begin{bmatrix} 3/16 \\ 11/16 \\ 3/16 \\ 11/16 \end{bmatrix}$,

$t^{(4)} = \begin{bmatrix} 7/32 \\ 23/32 \\ 7/32 \\ 23/32 \end{bmatrix}$, $\quad t^{(5)} = \begin{bmatrix} 15/64 \\ 47/64 \\ 15/64 \\ 47/64 \end{bmatrix}$, $\quad t^{(5)} - t = \begin{bmatrix} -1/64 \\ -1/64 \\ -1/64 \\ -1/64 \end{bmatrix}$

(d) for t_1: 5.4%; for t_2: -1.7%.

11.2 $\frac{1}{2}$

11.3 $t^{(1)} = \begin{bmatrix} \dfrac{3}{4} & \dfrac{5}{4} & \dfrac{2}{4} & \dfrac{5}{4} & \dfrac{4}{4} & \dfrac{2}{4} & \dfrac{5}{4} & \dfrac{4}{4} & \dfrac{3}{4} \end{bmatrix}^t$

$t^{(2)} = \begin{bmatrix} \dfrac{13}{16} & \dfrac{18}{16} & \dfrac{9}{16} & \dfrac{22}{16} & \dfrac{13}{16} & \dfrac{7}{16} & \dfrac{21}{16} & \dfrac{16}{16} & \dfrac{10}{16} \end{bmatrix}^t$

Chapter 12, page 162

12.2 $\begin{cases} a_n = \frac{1}{4} + (\frac{1}{2})^{n+1} (a_0 - c_0) \\ b_n = \frac{1}{2} \\ c_n = \frac{1}{4} - (\frac{1}{2})^{n+1} (a_0 - c_0) \end{cases}$ $n = 1, 2, \ldots$ $\begin{cases} a_n \to \frac{1}{4} \\ b_n = \frac{1}{2} \\ c_n \to \frac{1}{4} \end{cases}$ as $n \to \infty$

12.3
$$
\begin{cases}
a_{2n+1} = \dfrac{2}{3} + \dfrac{1}{6(4)^n} (2a_0 - b_0 - 4c_0) \\[4mm]
b_{2n+1} = \dfrac{1}{3} - \dfrac{1}{6(4)^n} (2a_0 - b_0 - 4c_0) \qquad n = 0, 1, 2, \ldots \\[4mm]
c_{2n+1} = 0
\end{cases}
$$

$$
\begin{cases}
a_{2n} = \dfrac{5}{12} + \dfrac{1}{6(4)^n} (2a_0 - b_0 - 4c_0) \\[4mm]
b_{2n} = \dfrac{1}{2} \qquad\qquad\qquad\qquad\qquad n = 1, 2, \ldots \\[4mm]
c_{2n} = \dfrac{1}{12} - \dfrac{1}{6(4)^n} (2a_0 - b_0 - 4c_0)
\end{cases}
$$

12.4 Eigenvalues: $\lambda_1 = 1$, $\lambda_2 = \tfrac{1}{2}$

Eigenvectors: $e_1 = \begin{bmatrix} 1 \\ 0 \end{bmatrix}$, $e_2 = \begin{bmatrix} 1 \\ -1 \end{bmatrix}$

12.5 12 generations; from 25% to .006%

12.6 $x^{(n)} =$

$$
\begin{bmatrix}
\dfrac{1}{2} + \dfrac{1}{2^{2n+3}} \left[(-3 - \sqrt{5})(1 + \sqrt{5})^{n+1} + (-3 + \sqrt{5})(1 - \sqrt{5})^{n+1} \right] \\[5mm]
\dfrac{1}{2^{2n+1}} \left[(1 + \sqrt{5})^{n+1} + (1 - \sqrt{5})^{n+1} \right] \\[5mm]
\dfrac{1}{2^{2n+1}} \left[(1 + \sqrt{5})^{n} + (1 - \sqrt{5})^{n} \right] \\[5mm]
\dfrac{1}{2^{2n+1}} \left[(1 + \sqrt{5})^{n} + (1 - \sqrt{5})^{n} \right] \\[5mm]
\dfrac{1}{2^{2n+1}} \left[(1 + \sqrt{5})^{n+1} + (1 - \sqrt{5})^{n+1} \right] \\[5mm]
\dfrac{1}{2} + \dfrac{1}{2^{2n+3}} \left[(-3 - \sqrt{5})(1 + \sqrt{5})^{n+1} + (-3 + \sqrt{5})(1 - \sqrt{5})^{n+1} \right]
\end{bmatrix}
$$

$$x^{(n)} \rightarrow \begin{bmatrix} \frac{1}{2} \\ 0 \\ 0 \\ 0 \\ 0 \\ \frac{1}{2} \end{bmatrix} \quad \text{as} \quad n \rightarrow \infty$$

12.8
$$\begin{bmatrix} 1 & 0 & 0 & 0 \\ 0 & 0 & 0 & 0 \\ 0 & 0 & 0 & 0 \\ 0 & 0 & 0 & 1 \end{bmatrix}$$

Chapter 13, page 178

13.1 (a) $\lambda_1 = 3/2$, $x_1 = \begin{bmatrix} 1 \\ 1/3 \end{bmatrix}$

(b) $x^{(1)} = \begin{bmatrix} 100 \\ 50 \end{bmatrix}$, $x^{(2)} = \begin{bmatrix} 175 \\ 50 \end{bmatrix}$, $x^{(3)} = \begin{bmatrix} 250 \\ 88 \end{bmatrix}$, $x^{(4)} = \begin{bmatrix} 382 \\ 125 \end{bmatrix}$

$x^{(5)} = \begin{bmatrix} 570 \\ 191 \end{bmatrix}$

(c) $x^{(6)} = Lx^{(5)} = \begin{bmatrix} 857 \\ 285 \end{bmatrix}$, $x^{(6)} \approx \lambda_1 x^{(5)} = \begin{bmatrix} 855 \\ 287 \end{bmatrix}$

13.7 2.375

13.8 1.49611

Chapter 14, page 192

14.1 (a) Yield = 33-1/3% of population; $x_1 = \begin{bmatrix} 1 \\ 1/3 \\ 1/18 \end{bmatrix}$

(b) Yield = 45.8% of population; $x_1 = \begin{bmatrix} 1 \\ 1/2 \\ 1/8 \end{bmatrix}$; harvest 57.9% of youngest age class.

14.2 $x_1 = \begin{bmatrix} 1.000 \\ .845 \\ .824 \\ .795 \\ .755 \\ .699 \\ .626 \\ .532 \\ 0 \\ 0 \\ 0 \\ 0 \end{bmatrix}$, $Lx_1 = \begin{bmatrix} 2.090 \\ .845 \\ .824 \\ .795 \\ .755 \\ .699 \\ .626 \\ .532 \\ .418 \\ 0 \\ 0 \\ 0 \end{bmatrix}$, $\dfrac{1.090 + .418}{7.584} = .199$

14.4 $h_I = (R - 1)/(a_I b_1 b_2 \cdots b_{I-1} + \cdots + a_n b_1 b_2 \cdots b_{n-1})$

14.5 $h_I = \dfrac{a_1 + a_2 b_1 + \cdots + a_{J-1} b_1 b_2 \cdots b_{J-2} - 1}{a_I b_1 b_2 \cdots b_{I-1} + \cdots + a_{J-1} b_1 b_2 \cdots b_{J-2}}$

Chapter 15, page 202

15.1 $y = -\dfrac{1}{2} + \dfrac{7}{2} x$

15.2 $y = \dfrac{2}{3} + \dfrac{1}{6} x$

15.3 $y = 2 + 5x - 3x^2$

15.4 $y = -5 + 3x - 4x^2 + 2x^3$

15.9 $y = 4 - .2x + .2x^2$; if $x = 12$ then $y = 30.4$ ($\$30.4$ thousand)

Chapter 16, page 217

16.1 $\dfrac{\pi^2}{3} + 4\cos t + \cos 2t + \dfrac{4}{9} \cos 3t$

16.2 $\dfrac{T^2}{3} + \dfrac{T^2}{\pi^2} \left(\cos \dfrac{2\pi}{T} t + \dfrac{1}{2^2} \cos \dfrac{4\pi}{T} t + \dfrac{1}{3^2} \cos \dfrac{6\pi}{T} t + \dfrac{1}{4^2} \cos \dfrac{8\pi}{T} t \right)$

$\qquad - \dfrac{T^2}{\pi} \left(\sin \dfrac{2\pi}{T} t + \dfrac{1}{2} \sin \dfrac{4\pi}{T} t + \dfrac{1}{3} \sin \dfrac{6\pi}{T} t + \dfrac{1}{4} \sin \dfrac{8\pi}{T} t \right)$

16.3 $\dfrac{1}{\pi} + \dfrac{1}{2} \sin t - \dfrac{2}{3\pi} \cos 2t - \dfrac{2}{15\pi} \cos 4t$

16.4 $\frac{4}{\pi}\left(\frac{1}{2} - \frac{1}{1\cdot 3}\cos t - \frac{1}{3\cdot 5}\cos 2t - \frac{1}{5\cdot 7}\cos 3t - \cdots - \frac{1}{(2n-1)(2n+1)}\cos nt\right)$

16.5 $\frac{T}{4} - \frac{8T}{\pi^2}\left(\frac{1}{2^2}\cos\frac{2\pi t}{T} + \frac{1}{6^2}\cos\frac{6\pi t}{T} + \frac{1}{10^2}\cos\frac{10\pi t}{T} + \cdots + \frac{1}{(2n)^2}\cos\frac{2n\pi t}{T}\right)$

Chapter 17, page 238

17.2 (a) $S(x) = -.76373(x-.4)^3 - .12244(x-.4)^2 + .93114(x-.4)$
$$+ .38942$$

(b) $S(.5) = .48055;$ Error $= .23\%$

17.3 (a) The cubic runout spline.

(b) $S(x) = 3x^3 - 2x^2 + 5x + 1$

17.4
$$S(x) = \begin{cases} -.00000042(x+10)^3 & + .000214(x+10) + .99815, & -10 \le x \le 0 \\ .00000024(\ x\ )^3 - .0000126(\ x\ )^2 + .000088(\ x\ ) + .99987, & 0 \le x \le 10 \\ -.00000004(x-10)^3 - .0000054(x-10)^2 - .000092(x-10) + .99973, & 10 \le x \le 20 \\ .00000022(x-20)^3 - .0000066(x-20)^2 - .000212(x-20) + .99823, & 20 \le x \le 30 \end{cases}$$

Maximum at $(x, S(x)) = (3.93, 1.00004)$.

17.5
$$S(x) = \begin{cases} .00000009(x+10)^3 - .0000121(x+10)^2 + .000282(x+10) + .99815, & -10 \le x \le 0 \\ .00000009(\ x\ )^3 - .0000093(\ x\ )^2 + .000070(\ x\ ) + .99987, & 0 \le x \le 10 \\ .00000004(x-10)^3 - .0000066(x-10)^2 - .000087(x-10) + .99973, & 10 \le x \le 20 \\ .00000004(x-20)^3 - .0000053(x-20)^2 - .000207(x-20) + .99823, & 20 \le x \le 30 \end{cases}$$

Maximum at $(x, S(x)) = (4.00, 1.00001)$.

17.6 (b)

$$\begin{bmatrix} 4 & 1 & 0 & 0 & \cdots & 0 & 0 & 0 & 1 \\ 1 & 4 & 1 & 0 & \cdots & 0 & 0 & 0 & 0 \\ 0 & 1 & 4 & 1 & \cdots & 0 & 0 & 0 & 0 \\ \vdots & \vdots & \vdots & \vdots & & \vdots & \vdots & \vdots & \vdots \\ 0 & 0 & 0 & 0 & \cdots & 0 & 1 & 4 & 1 \\ 1 & 0 & 0 & 0 & \cdots & 0 & 0 & 1 & 4 \end{bmatrix} \begin{bmatrix} M_1 \\ M_2 \\ M_3 \\ \vdots \\ M_{n-2} \\ M_{n-1} \end{bmatrix} = \frac{6}{h^2} \begin{bmatrix} y_{n-1} - y_1 + y_2 \\ y_1 - 2y_2 + y_3 \\ y_2 - 2y_3 + y_4 \\ \vdots \\ y_{n-3} - 2y_{n-2} + y_{n-1} \\ y_{n-2} - 2y_{n-1} + y_1 \end{bmatrix}$$

$$
\begin{bmatrix}
2 & 1 & 0 & 0 & \cdots & 0 & 0 & 0 & 0 \\
1 & 4 & 1 & 0 & \cdots & 0 & 0 & 0 & 0 \\
0 & 1 & 4 & 1 & \cdots & 0 & 0 & 0 & 0 \\
\vdots & \vdots & \vdots & \vdots & & \vdots & \vdots & \vdots & \vdots \\
0 & 0 & 0 & 0 & \cdots & 0 & 1 & 4 & 1 \\
0 & 0 & 0 & 0 & \cdots & 0 & 0 & 1 & 2
\end{bmatrix}
\begin{bmatrix}
M_1 \\ M_2 \\ M_3 \\ \vdots \\ M_{n-1} \\ M_n
\end{bmatrix}
= \frac{6}{h^2}
\begin{bmatrix}
-hy_1' - y_1 + y_2 \\
y_1 - 2y_2 + y_3 \\
y_2 - 2y_3 + y_4 \\
\vdots \\
y_{n-2} - 2y_{n-1} + y_n \\
y_{n-1} - y_n + hy_n'
\end{bmatrix}
$$

17.7 (b)

Chapter 18, page 257

18.1 (a) *GIYUOKEVBH*   (b) *SFANEFZWJH*

18.2 (a) $A^{-1} = \begin{bmatrix} 12 & 7 \\ 23 & 15 \end{bmatrix}$ (b) Not invertible.   (c) $A^{-1} = \begin{bmatrix} 1 & 19 \\ 23 & 24 \end{bmatrix}$

(d) Not invertible.   (e) Not invertible.   (f) $A^{-1} = \begin{bmatrix} 15 & 12 \\ 21 & 5 \end{bmatrix}$

18.3 *WE LOVE MATH*

18.4 Deciphering matrix $= \begin{bmatrix} 7 & 15 \\ 6 & 5 \end{bmatrix}$;   enciphering matrix $= \begin{bmatrix} 7 & 5 \\ 2 & 15 \end{bmatrix}$

18.5 *THEY SPLIT THE ATOM*

18.6 *I HAVE COME TO BURY CAESAR*

18.7 A is invertible modulo 29 if and only if $\det(A) \neq 0$ (mod 29).

Chapter 19, page 281

19.1 (c) $x_3^* = (31/22, \ 27/22)$

19.2 (a) $x_3^{(1)} = (1.40000, 1.20000)$ (c) $x_3^{(1)} = (9.55000, 25.65000)$

$x_3^{(2)} = (1.41000, 1.23000)$ $x_3^{(2)} = (\ .59500, -1.21500)$

$x_3^{(3)} = (1.40900, 1.22700)$ $x_3^{(3)} = (1.49050, \ 1.47150)$

$x_3^{(4)} = (1.40910, 1.22730)$ $x_3^{(4)} = (1.40095, \ 1.20285)$

$$x_3^{(5)} = (1.40909, 1.22727) \qquad x_3^{(5)} = (1.40991, 1.22972)$$

$$x_3^{(6)} = (1.40909, 1.22727) \qquad x_3^{(6)} = (1.40901, 1.22703)$$

(b) Same as part (a).

19.4 $x_1^* = (1, 1)$, $x_2^* = (2, 0)$, $x_3^* = (1, 1)$

19.7

$$x_7 + x_8 + x_9 = 13.00$$
$$x_4 + x_5 + x_6 = 15.00$$
$$x_1 + x_2 + x_3 = 8.00$$
$$.82843(x_6 + x_8) + .58579x_9 = 14.79$$
$$1.41421(x_3 + x_5 + x_7) = 14.31$$
$$.82843(x_2 + x_4) + .58579x_1 = 3.81$$
$$x_3 + x_6 + x_9 = 18.00$$
$$x_2 + x_5 + x_8 = 12.00$$
$$x_1 + x_4 + x_7 = 6.00$$
$$.82843(x_2 + x_6) + .58579x_3 = 10.51$$
$$1.41421(x_1 + x_5 + x_9) = 16.13$$
$$.82843(x_4 + x_8) + .58579x_7 = 7.04$$

19.8 (a)

$$x_7 + x_8 + x_9 = 13.00$$
$$x_4 + x_5 + x_6 = 15.00$$
$$x_1 + x_2 + x_3 = 8.00$$
$$.04289(x_3 + x_5 + x_7) + .75000(x_6 + x_8) + .61396x_9 = 14.79$$
$$.91421(x_3 + x_5 + x_7) + .25000(x_2 + x_4 + x_6 + x_8) = 14.31$$
$$.04289(x_3 + x_5 + x_7) + .75000(x_2 + x_4) + .61396x_1 = 3.81$$
$$x_3 + x_6 + x_9 = 18.00$$
$$x_2 + x_5 + x_8 = 12.00$$
$$x_1 + x_4 + x_7 = 6.00$$

$$.04289\,(x_1 + x_5 + x_9) + .75000\,(x_2 + x_6) + .61396x_3 \;=\; 10.51$$

$$.91421\,(x_1 + x_5 + x_9) + .25000\,(x_2 + x_4 + x_6 + x_8) \;=\; 16.13$$

$$.04289\,(x_1 + x_5 + x_9) + .75000\,(x_4 + x_8) + .61396x_7 \;=\; 7.04$$

Chapter 20, page 299

20.1 $x_1 = 2$, $x_2 = 2/3$; maximum value of $z = 22/3$

20.2 no feasible solutions

20.3 unbounded solution

20.4 invest \$6,000 in bond A and \$4,000 in bond B; the annual yield is \$880.

20.5 7/9 cup of milk, 25/18 ounces of corn flakes; minimum cost = 335/18 = 18.6¢

20.6 550 containers from company A and 300 containers from company B; maximum shipping charges = \$2,110

20.7 925 containers from company A and no containers from company B; maximum shipping charges = \$2,312.50

20.8 .4 pound of ingredient A and 2.4 pounds of ingredient B; minimum cost = 24.8¢

Chapter 21, page 319

21.1
$$\text{Maximize} \quad z' = -2x_1 - 5x_2$$

subject to

$$
\begin{aligned}
3x_1 - 6x_2 + x_3 &&&&&= 2\\
x_1 + x_2 &&+ x_4 &&&= 3\\
-x_1 &&&&+ x_5 &= -6\\
x_2 &&&&&+ x_6 = 5
\end{aligned}
$$

and

$$x_1,\, x_2,\, x_3,\, x_4,\, x_5,\, x_6 \geq 0$$

21.2
$$\text{Maximize} \quad z = -3x_1 + x_2 + x_3$$

subject to

$$3x_1 - 5x_2 + x_3 = 3$$
$$-2x_1 - x_2 + x_4 = 2$$
$$-x_1 + x_5 = -5$$
$$-x_2 + x_6 = -2$$

and

$$x_1, \ x_2, \ x_3, \ x_4, \ x_5, \ x_6 \geq 0$$

21.3 (b) $\begin{bmatrix} 1/2 \\ 3/2 \\ 0 \\ 0 \end{bmatrix}, \quad \begin{bmatrix} 2 \\ 0 \\ 0 \\ -3 \end{bmatrix}, \quad \begin{bmatrix} 0 \\ 3/2 \\ 1/4 \\ 0 \end{bmatrix}, \quad \begin{bmatrix} 0 \\ 2 \\ 0 \\ 1 \end{bmatrix}, \quad \begin{bmatrix} 0 \\ 0 \\ 1 \\ -3 \end{bmatrix}$

(c) $\mathbf{x}_1 = \begin{bmatrix} 1/2 \\ 3/2 \\ 0 \\ 0 \end{bmatrix}, \qquad \mathbf{x}_2 = \begin{bmatrix} 0 \\ 3/2 \\ 1/4 \\ 0 \end{bmatrix}, \qquad \mathbf{x}_3 = \begin{bmatrix} 0 \\ 2 \\ 0 \\ 1 \end{bmatrix}$

(d) $x_1 = 0, \ x_2 = 2, \ x_3 = 0, \ x_4 = 1; \quad z = 1$

(e)

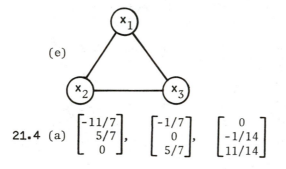

21.4 (a) $\begin{bmatrix} -11/7 \\ 5/7 \\ 0 \end{bmatrix}, \quad \begin{bmatrix} -1/7 \\ 0 \\ 5/7 \end{bmatrix}, \quad \begin{bmatrix} 0 \\ -1/14 \\ 11/14 \end{bmatrix}$

21.5 (a)   Maximize $z = 3x_1 + 2x_2 - x_3 + 0x_4 + 0x_5$

subject to

$$2x_1 + 3x_2 + x_3 + x_4 = 4$$
$$x_1 + 2x_2 + 3x_3 + x_5 = 5$$

and

$$x_1, \ x_2, \ x_3, \ x_4, \ x_5 \geq 0$$

(b) $\begin{bmatrix} -7 \\ 6 \\ 0 \\ 0 \\ 0 \end{bmatrix}, \begin{bmatrix} 7/5 \\ 0 \\ 6/5 \\ 0 \\ 0 \end{bmatrix}, \begin{bmatrix} 5 \\ 0 \\ 0 \\ -6 \\ 0 \end{bmatrix}, \begin{bmatrix} 2 \\ 0 \\ 0 \\ 0 \\ 3 \end{bmatrix}, \begin{bmatrix} 0 \\ 1 \\ 1 \\ 0 \\ 0 \end{bmatrix}$

$\begin{bmatrix} 0 \\ 5/2 \\ 0 \\ -7/2 \\ 0 \end{bmatrix}, \begin{bmatrix} 0 \\ 4/3 \\ 0 \\ 0 \\ 7/3 \end{bmatrix}, \begin{bmatrix} 0 \\ 0 \\ 5/3 \\ 7/3 \\ 0 \end{bmatrix}, \begin{bmatrix} 0 \\ 0 \\ 4 \\ 0 \\ -7 \end{bmatrix}, \begin{bmatrix} 0 \\ 0 \\ 0 \\ 4 \\ 5 \end{bmatrix}$

(c) $\mathbf{x}_1 = \begin{bmatrix} 7/5 \\ 0 \\ 6/5 \\ 0 \\ 0 \end{bmatrix}, \quad \mathbf{x}_2 = \begin{bmatrix} 2 \\ 0 \\ 0 \\ 0 \\ 3 \end{bmatrix}, \quad \mathbf{x}_3 = \begin{bmatrix} 0 \\ 1 \\ 1 \\ 0 \\ 0 \end{bmatrix},$

$\mathbf{x}_4 = \begin{bmatrix} 0 \\ 4/3 \\ 0 \\ 0 \\ 7/3 \end{bmatrix}, \quad \mathbf{x}_5 = \begin{bmatrix} 0 \\ 0 \\ 5/3 \\ 7/3 \\ 0 \end{bmatrix}, \quad \mathbf{x}_6 = \begin{bmatrix} 0 \\ 0 \\ 0 \\ 4 \\ 5 \end{bmatrix}$

(d) $x_1 = 2$, $x_2 = 0$, $x_3 = 0$, $x_4 = 0$, $x_5 = 3$; $z = 6$

(e)

(f) $x_1 = 2$, $x_2 = 0$, $x_3 = 0$; $z = 6$

21.6 (a) Maximize $z' = -2x_1 + 3x_2 - x_3 + 0x_4$

subject to

$$x_1 - 2x_2 + 3x_3 + x_4 = 5$$
$$2x_1 + x_2 - 2x_3 = 2$$

and

$$x_1, x_2, x_3, x_4 \geq 0$$

(b) $\begin{bmatrix} 9/5 \\ -8/5 \\ 0 \\ 0 \end{bmatrix}$, $\begin{bmatrix} 2 \\ 0 \\ 1 \\ 0 \end{bmatrix}$, $\begin{bmatrix} 1 \\ 0 \\ 0 \\ 4 \end{bmatrix}$, $\begin{bmatrix} 0 \\ -16 \\ -9 \\ 0 \end{bmatrix}$, $\begin{bmatrix} 0 \\ 2 \\ 0 \\ 9 \end{bmatrix}$, $\begin{bmatrix} 0 \\ 0 \\ -1 \\ 8 \end{bmatrix}$

(c) $x_1 = \begin{bmatrix} 2 \\ 0 \\ 1 \\ 0 \end{bmatrix}$, $x_2 = \begin{bmatrix} 1 \\ 0 \\ 0 \\ 4 \end{bmatrix}$, $x_3 = \begin{bmatrix} 0 \\ 2 \\ 0 \\ 9 \end{bmatrix}$

(d) $x_1 = 0$, $x_2 = 2$, $x_3 = 0$, $x_4 = 9$; $z' = 6$

(e)

(f) $x_1 = 0$, $x_2 = 2$, $x_3 = 0$; $z = -6$

Chapter 22, page 340

22.1 $x_1 = 0$, $x_2 = 3/2$; $z = 6$

22.2 $x_1 = 2/3$, $x_2 = 1$; $z = 7/3$

22.3 $x_1 = 0$, $x_2 = 1/2$, $x_3 = 9/2$; $z = 26$

22.4 $x_1 = 22/13$, $x_2 = 6/13$, $x_3 = 0$; $z = 50/13$

22.5 $x_1 = 16/5$, $x_2 = 0$, $x_3 = 0$, $x_4 = 2/5$; $z = 10$

22.6 $x_1 = 0$, $x_2 = 13$, $x_3 = 34$, $x_4 = 0$; $z = 76$

22.7 $x_1 = 260$, $x_2 = 0$; $z = 520$

22.8 $x_1 = 2$, $x_2 = 2/3$; $z = 22/3$

22.9 $x_1 = 550$, $x_2 = 300$; $z = 2110$

22.10 $x_1 = 925$, $x_2 = 0$; $z = 2312.50$